TELEPHONY

*A MANUAL OF THE
DESIGN, CONSTRUCTION, AND OPERATION
OF TELEPHONE EXCHANGES*

IN SIX PARTS

Part I.

THE LOCATION OF CENTRAL OFFICES

WITH 33 ILLUSTRATIONS

BY

ARTHUR VAUGHAN ABBOTT, C. E.

NEW YORK
McGRAW PUBLISHING COMPANY.
1903

PREFACE.

IT is veraciously reported that a quarter of a century ago, when a well-known New York capitalist was asked to invest in some Bell Telephone stock, he gave for his refusal the reason that the American District Telegraph boys were amply sufficient, and a Telephone enterprise would surely fail, as there was no room for an additional company in the business of transmitting messages. To-day the A. D. T. is a byword, while Telephony, ranking as one of the foremost of modern industries, bravely holds its own in the electrical quintet of the Telegraph, the Telephone, the Street Railway, Electric Lighting, and Power Transmission. As a factor in modern civilization it is a peer of any electrical industry, while from a financial standpoint it is, to say the least, a lusty third, with a strong probability that ere long it may lead the list.

There is so much literature descriptive of the telephone as an instrument for the transmission of speech, and explanatory of its method of operation, that a voluminous apology would be needed for any addition thereto. But telephone plants are now assuming such proportions that the questions of *economic design* and *operation* become paramount, and the "*Telephone Engineer*" is coequal with his confrere of the Electric Light, the Street Railway, or

the Power Transmission Plant. From the engineering
point of view little has been written. Largely this is due
to the newness of the art, for in most cases telephone
plants, like Topsy, "simply grew"; but accumulated experience is rapidly demonstrating that certain underlying
principles of design are absolutely essential to success.
As the business extends competitive companies arise, and
the possible margin of profit decreases, until only those
that are installed and managed with the utmost economy
can become attractive investments. Before capital can be
secured, a carefully worked-out design and estimate for
each plant must be prepared; and to be successful such
design must be based upon principles which have received
the refinement of the furnace of experience. After construction is completed, operation must likewise be carried
on upon a basic foundation of knowledge that has stood
the test of time, conjointly with such a perception of the
needs of subscribers as will enable the manager to, at all
times, meet the requirements of a constantly progressing
public. As Telephony is exceedingly young, the experience of yesterday can only indicate the *direction* of the
lines along which the installation and operation of the
plant of to-morrow should proceed. It is the endeavor of
the author to collect in this series the concensus of what
appears to be at present the best prevailing practice, in
the hope that it may at least serve as a guide-post along
the telephonic road.

The end and aim of design in Telephony is the installation of the best plant for the least money. At the outset
the telephone engineer is confronted with a fixed condition in the location of subscribers, distributed over a

definite area; and the problem is to produce such a plant as will serve them best, for the least expenditure. The central office site must be chosen; routes for lines selected; a wire plant created; sub-stations installed, and connected to a switch board in the central office. As the sub-stations are fixed in position, and as wire-plant routes must follow streets already in existence, the location of the central office is the chief variable; and its selection is of first consideration, for upon its situation the quantity and arrangement of the wire plant will very largely depend.

The present presentation, in Vol. I., " *The Location of the Central Office*," commences with the rules which experience has shown as advisable guides in the selection of the office site.

ARTHUR VAUGHAN ABBOTT.

New York, March, 1903.

CONTENTS.

PART I.

LIST OF TABLES.

PART I.

LIST OF ILLUSTRATIONS.

PART I.

INTRODUCTION.

TELEPHONIC design as one department of the great field of electrical engineering has had few exponents, and so far as the writer is aware, there has been no connected exposition of the principles which govern the mutual relations of the various parts of a plant necessary to telephonic service on a large scale, in such a manner as to enable the engineer to so proportion the quantity and quality of the essential apparatus as to produce an installation that will give the best service with a minimum initial investment, and aid the operating manager to so prosecute the business as to insure satisfactory service with the least annual expense.

The reasons for this absence of literature are not far to seek. Telephony as an art is less than a quarter of a century old; and those engaged therein have had to feel their way step by step, usually so handicapped by the demands of a business that was increasing too fast for experience to overtake, as to lack the time and opportunity to record the lessons that actual practice inculcated. Then, for the best part of its present lifetime, the telephone has been under the control of a management that, while it has usually prosecuted the development of the art in a most careful and praiseworthy manner, ha steadily indulged in what appears to the author the mi taken policy of repressing free discussion of the subjec This state of affairs is fast disappearing. With the ind pendent movement, exchanges are springing up in eac town all over the country, and each State is organizin mutual telephone associations at whose frequent meeting a full and frank interchange of views and experience exchanged, that the technical press immediately places the service of those interested, thus imitating the broade and more liberal methods of other electrical interests.

During the experience of a decade in the engineerin field of telephony, the writer has had an opportunity t watch the operation of what appear to be some of th fundamental principles of the economic design and mai agement of telephone exchanges. At best these observi tions can only cover a small fraction of even preser telephonic experience, and deductions therefrom must k accepted with the greatest caution, and be subject t extreme modifications, both in place and in time. Wit such restrictions, these notes are now offered for the coi sideration of those who are, perhaps, in an earlier stage c experience, in the hope that in some cases, at least, the may serve as a guide of what not to do.

TELEPHONY.

CHAPTER I.

THE FACTORS IN THE PROBLEM OF DESIGN.

An attentive observer at the Philadelphia Centennial Exposition could have noticed a fine wire extending along one of the galleries of Machinery Hall. The ends of the wire terminated in curious trumpet-shaped pieces of apparatus, placed in little inclosures situated at opposite sides of the building, wherewith a careful listener, located at one end of the wire, aided by acute hearing, and the judicious use of the imagination, might understand articulated sentences shouted into the apparatus at the other end. Such was the first public exhibition of the telephone.

By means of the invention of Professor Bell, the long-sought electrical transmission of speech became an accomplished fact; and during the quarter of a century which has since elapsed, electrical arachnida have been so industriously spinning a copper net over the continent, that in the United States alone, there are approximately 1,000,000 miles of telephone wire, bringing into daily speaking relations about 2,500,000 of subscribers; that with 3,000,000 telephones, annually exchange nearly 4,000,000,000 messages, while for the prosecution of this business some tens of thousands of employes are needed.

Though Professor Bell's discovery richly merits the distinction of being the keystone of the stupendous telephonic edifice that has since been reared, yet the present value of the telephone and its influence in business and social intercourse depend quite as much on other and auxiliary apparatus that, voussoir-like, supports and sustains the key, and is equally necessary to complete the magnificent structure. Bell's magneto telephone supplied an instrument for mutually interconverting sound and electricity — absolutely nothing more. But to render inter-communication practical, calling signals were needed; for, although the telephone could talk, its voice was too feeble to gain notice, and such a contrivance was required as would at once attract the attention of those in the vicinity of the instrument, without requiring watching, and also one that would be perfectly reliable and need no attention to keep in constant repair. Without an exception, all forms of signals for subscribers, excepting the audible one, have proved to be failures; and while multitudes of devices involving the use of the receiver or an electro-magnetic tuning fork arranged to emit a musical note have been tried, all have been relegated to obscurity save the familiar call bell.

In early days the old-fashioned vibrating bell, with battery and push-button, was immediately at hand, and gained a widespread introduction; but the battery was inconvenient to maintain and the bell contacts got out of order, so the magneto bell, containing no contacts, operated by a small electric hand dynamo, displaced in the race for survival the electro-chemical signal.

Experience soon proved that the magneto telephone was exceedingly inefficient in the transmission of speech, as it could only be used over short lines, and even then necessitated the expenditure of much lung power at the sending-end. No sooner did Bell's instrument appear than a host of experimenters set to work, and presently the microphonic inventions of Berliner, Gray, Hughes, Edison, and Blake completely displaced the Bell telephone for all purposes of transmitting speech. The Bell type as a receiver, however, still holds undisputed sway in almost the identical original form. So by the schooling of experience a telephonic outfit for the subscriber has been refined to a magneto signal, a battery transmitter, and magneto receiver.

In addition to talking and signaling apparatus for the telephone user, a path over which electrical energy could travel between correspondents was required; for in early telephone days Hertzian waves were entirely unknown, and even to-day their telephonic possibilities are of the remotest. Hence the unsightly aerial lines which grew like mushrooms, exhibiting with their crooked knotted poles, topheavy with almost impenetrable cobwebs of wires, and loaded with wrecks of innumerable kites, many other cryptogamic resemblances. Gradually, however, the open-wire line evolved into the aerial cable, and from the aerial cable developed the much more satisfactory underground system that now only betrays its presence by periodically recurring upheavals of the streets, and a chronic disease of city pavements. In the beginning each person was compelled to have a separate line extending to every other with whom conversation was desired, though with the first handful of telephones installed the absurdity of this method, from both the standpoint of expense and intricacy, was demonstrated. To reduce the number of wires to a minimum, only a single one was strung for each pair of communicants, Mother Earth being used as a common return; but like all garrulous gossips, she caused trouble by failing to hold her tongue, creating no end of domestic disturbance by surreptitiously informing Mrs. X. what Mrs. Y. had ordered for dinner. Then when the electrical railway and electric light made their appearance conversation of all kinds was so interjected with such a hissing and spitting that telephonists were fain to immediately convert each line into a complete metallic circuit.

As an escape from the impracticability of a wire for each pair of talkers, came the germ of the modern exchange in the idea of bringing all lines to some central point. At first the attendant carefully and laboriously wrote out the message that the calling party wished to transmit, summoned the desired correspondent and yelled the message into a plain magneto-telephone (for in those days there were no solid backs), this performance requiring a sufficient expenditure of lung power to enable the recipient of the message to receive the same independently of the telephone, provided he happened to be within a few blocks of the office. But this plan limited the use of the telephone to messages which were purely mandatory; and as people wanted to converse with each other, some means of interconnecting different lines was needed. The old telegraphic device of interlaced brass strips was the next step in advance, by means of which, with the aid of plugs fitted into holes in the strips, any pair of lines could be promptly joined together. This plan worked with fair success until the number of subscribers so increased as to require a greater space for the strips than could be covered by the arms of four operators seated at a square table. At this point the invention of the spring-jack and flexible cord terminating in a metallic plug, whereby any two jacks each attached to subscriber's lines could be quickly and surely connected, afforded immediate and seemingly permanent relief. But the increase in business has been so rapid that the ingenuity of inventors has been taxed to the utmost to reduce the jack to its least dimensions, in order that all the subscribers brought to one office could be placed within an arm's reach of the operator. Hand in hand with the evolution of the spring-jack and connecting cords, the development of central station signals has taken place, for switchboard requirements are very different from those of the subscriber. At the office an attendant is supposed to be constantly on duty to watch signals; but it was essential that the party calling should be plainly identified — a function entirely unnecessary at the subscriber's premises. The call-bell was tried, but found too bulky and indistinctive. The electromagnetic annunciator in the form of the familiar hotel apparatus, equipped with a falling shutter, or drop, whereby a number painted thereon could be displayed at the subscriber's pleasure, reigned supreme for more than a decade; but the decreasing size of spring-jacks with which the signal must be associated, and the demands for service that should be surer and more prompt, have consigned the drop to the scrap heap, and substituted therefor a miniature incandescent lamp, that is now universal in modern switchboards. So, from a little table in a small room in Devonshire Street, Boston, to which a half-dozen magneto telephones were connected by iron wires strung in streets, the modern telephone system of the country has evolved, counting its subscribers by millions, and talking with ease over half the continent.

Every telephone plant, therefore, may be structurally divided into three separate parts, that perform entirely different and separate functions, the sum of which is necessary to the prosecution of the business. These parts are :

First. — *The subscriber's station*, containing apparatus to enable the user to transmit signals to and from the office and to transmit and receive conversation.

Second. — *The line*, comprising the necessary wire plant to supply electrical communication between each subscriber and the central office, and between different central offices.

Third. — *The central office*, comprising apparatus capable of performing the following functions :

First. — (a) To permit any subscriber to call the attendant at all times.

(b) To permit any subscriber to give necessary instructions to the attendant.

(c) To permit the attendant to ascertain whether any subscriber is engaged in conversation without causing interruption.

(d) To permit the attendant to call any disengaged subscriber.

(e) To permit the attendant to connect together any two parties for conversation.

(f) To notify the attendant at once when conversation is completed.

Second. —To enable attendants to perform the same functions between different offices.

The *Economic Design* of a *Telephone Exchange* is such a selection and arrangement of apparatus and appliances for each of the factors thus defined, and such a mutual adjustment of one to the other, as to enable the particular plant under consideration to give the best service with the least initial investment. The *Economic Management* of a *Telephone Exchange* is such a regular manipulation of the plant as shall secure the best service with the least *continued* annual expense.

In the consideration of the three telephonic factors taken in the preceding order, it is most convenient to commence with the central office as a center and proceed centrifugally outward to the subscriber. Telephonists, however, have a language of their own, partly composed of words and phrases peculiar to the art, and partly of expressions current in electrical and mechanical engineering, to which a specialized meaning, different from that ordinarily conveyed, is attached, and even between experts the same terms are used with varying meaning. It is well, therefore, before proceeding to the consideration of central station design, to rehearse the more common telephonic definitions in order that in future, paragraphic brevity may be combined with perfect definiteness of statement.

CHAPTER II.

DEFINITIONS.

As a preliminary to entering into the details of the subject, the following definitions of terms that will constantly occur are given:

GENERAL OR FUNDAMENTAL TERMS.

a. *Central Office, Central Station* or *Central.* — The building, or office space, switchboard, and all auxiliary apparatus, excluding cable entrance and cable terminals, required for placing subscribers in talking relations with each other, without the use of trunk lines extending outside the building in which the switchboard is located.

b. *Exchange.* — A collection of central offices connected by trunk lines over which any subscriber tributary to any of these offices may talk without paying toll charges, often used synonymously with central office. An exchange must have one central office, and it may have several.

c. *Territorial Exchange.* — A collection of exchanges and connecting toll-lines under the management of one corporation.

d. *Wire Plant* or *Line Plant.* — All of the installation required to provide electrical connection between the outside walls of sub-stations (see *Sub-Station*), and the main distributing boards (see *Main Distributing Board*) of the Central offices to which the sub-stations are tributary, and in addition the plant connecting the *Main Distributing Boards*, of the various central offices in a given exchange.

e. *Sub-Stations.* — The term " Sub-Station " designates either the place at which the subscriber's outfit is located, or generically all the apparatus required for the equipment of a subscriber.

f. *Branch Exchange.* — There are many cases of isolated groups of subscribers who desire to talk frequently among themselves, and to have occasional connection with other exchange subscribers. To pass all the messages that such a group originates through the central office would require an unnecessary outlay, both in apparatus and service. Hence it is customary to place such subscribers in a small switchboard by themselves, connecting them to the central office by one or more trunk lines. Such an installation is called a *Branch Exchange.*

g. *Bridge.* — A shunt; electrical circuits, or pieces of apparatus carrying electrical currents in parallel are said to be bridged or in bridge.

h. *Message,* often called *call* or *connection,* sometimes a *switch.* — The act of signaling the central office, also the conversation which takes place between two subscribers. Messages are divided into (1) originating messages, or those which are made by subscribers in signaling the office to which they are attached, and (2) trunk messages, or such messages as proceed from the office in which they originate over trunk lines through some other office to the called subscriber.

i. *Noise, Induction, Cross-Talk.* — Any line on which any sound can be distinguished other than the conversation of parties at the ends thereof, is said to be *noisy* or afflicted with *induction*. When conversation passing over a neighboring line can be distinguished there is said to be *cross-talk.*

j. *Private Exchange.* — A branch exchange without trunk lines to a central office.

k. *Repeating Coil.* — Two coils of wire so wound on the same core that electric impulses in one will excite corresponding impulses in the other; a transformer. Repeating coils may be step-up, even, or step-down, depending on the mutual relations of the primary and secondary windings.

l. *Retardation* or *Retardation Coil.* — An inductive resistance.

m. *Telephonic Center.* — That point within any group of subscribers from which the sum of the distances, measured along streets, to all sub-stations, is minimum.

n. *Trouble.* — When any piece of apparatus is out of order, or a circuit does not operate in a normal manner, it is said to be in *trouble.* Employes, whose business it is to remedy such defects, are called *Trouble Men,* and the difficulty when removed is said to be *Cleared.*

SUBSIDIARY TERMS.

Central Office Equipment. — The entire outfit, excluding building, cable entrance, and cable terminals needed for a central office, usually divided into four parts: (1) The switchboard, (2) the distributing boards, (3) the power plant, (4) the operator's furniture.

1. THE SWITCHBOARD.

The Switchboard. — The switchboard is that portion of a central office equipment provided to enable operators to answer subscribers' signals, connect them together in talking relations, and disconnect lines when conversation is complete.

Types of Board. — There are two distinctive fundamental types of switchboard in use. One, the *Multiple Board*, in which on each subscriber's line, the jacks (see Jack) are repeated or *Multiplied* a sufficient number of times to place a jack on each line within the reach of every operator that can sit at the board ; the other, the *Transfer Board*, in which there is only one jack to each line, and every operator who wishes to connect a subscriber whose jack is not in front of her must *transfer* the message to some other operator before whom the desired subscriber's jack does appear. These two types shade into each other in all degrees.

There are also three other distinctive switchboard features leading to a subordinate classification under the preceding general divisions. These three features involve the methods whereby electrical energy is supplied for signaling and talking. (a) *Magneto Boards* are those in which the calling and clearing-out signals (see Signals) are actuated by current obtained from a small magneto generator driven manually by the subscriber, and electricity for talking is supplied by a local battery at each substation. (b) *Automatic Signal Boards* are those in which the calling and clearing-out signals are given automatically by the removal and replacement of the receiver (see Receiver) on the hook (see Hook), electricity therefor being supplied from a battery at the central office. Under this division also fall a small number of boards, tending to become obsolete, wherein the call signal is made by pushing a button, and the clearing out is automatic on the replacement of the receiver, battery being supplied from the central office to actuate the signals only. (c) *Common Battery Boards* are those in which *all* electricity, both for automatic signaling and talking, is supplied to all subscribers from one *Common Battery* located at the central office. Common Battery Boards are frequently called Relay Boards, or Central Energy Systems, and automatic signal boards are also, though erroneously, so termed. It is obvious that these types of board may be built on either the multiple or transfer system.

Sub-Divisions of Switchboards; A & B Boards. — The *Switchboard* in every exchange having more than one central office is divided into two parts, one devoted to receiving calls from the subscribers connected directly to the office under consideration, and in making connection between such subscribers and others either located in the same or other offices. This part of the switchboard is called the *A Board*, the *Subscribers' Board*, the *Answering Board* or the *Originating Board*. The other portion of the switchboard is devoted to receiving calls made by subscribers connected to *other offices only*, and in making connections only to subscribers in the office under consideration. This portion of the board is called the *B Board*, the *Trunking Board*, or the *Connecting Board*. Where there is only one central office, no B boards exist. The construction of these two parts of the switchboard is, in some features, radically different.

Position. — Every switchboard may be considered to be made up of a number of exactly similar yet entirely distinct units, each unit being that amount of apparatus necessary to the vocations of *one single operator.* Such a collection of apparatus is called a *Position.* As the functions of an operator at an " A " Board are different from those at a " B " Board, the " A " positions will be differently equipped from the " B " positions, so it is usual to speak of any switchboard as consisting of so many " A " positions and so many " B " positions; and as each position of each kind is similar to all the others of the same kind, the total cost of the board is easily obtained by multiplying the cost of each kind of a position by the number of each.

" A " Position. — Each " A " position may be divided into three parts. (1) The Subscriber's Line Equipment. This is usually understood to embrace a protector (see Protection), a pair of terminals (see Terminal) in the main and intermediate boards (see Distributing Board), an answering jack and signal (see Signal), together with all connecting cable and other auxiliary apparatus. In magneto boards the signal is a drop (see Drop), but in automatic signal and common battery boards, it is a small incandescent lamp, operated by a relay (see Relay), and the subscriber's line equipment includes the lamp and socket, relay, and cut-off, or controlling relay for extinguishing the lamp when the operator answers, together with a unit portion of the relay racks and connecting cables.

(2) *The Multiple.* The *Multiple* or *Multiple Jacks* comprise the entire outfit of jacks needed to enable each

operator to transact all the business that comes to her. The Multiple Jacks are divided into two parts, — the *Subscribers' Multiple*, or the jacks that are attached to subscribers' lines, and the *Out-Trunk Multiple*, or the jacks that are provided to enable each " A " operator to have access to all of the trunk lines (see Trunk Line) that extend to other offices.

(3) *The Operator's Equipment.*—Each operator must be provided with apparatus wherewith two lines may be connected together, must be able to receive instructions from the calling subscriber, and be able to ring the party called for. To connect lines together it is usual to provide a pair of plugs (see Plug), fitting the spring jacks that are connected by a flexible cord (see Cord); by placing one of the pair (called the answering plug) into the jack of the subscriber calling, and the other (called the connecting plug) into the jack of the one called for, the two lines are connected together. Each pair of cords is supplied with a switch or key (see Key), so constructed that when the handle is moved in one direction the receiver and transmitter supplied to the operator are connected to the line, and when in the other the operator sends ringing current over the line of the called subscriber. In magneto boards, a drop placed across the cords notifies the operator when conversation is complete; while in Automatic Signal and Common Battery boards, a pair of lamps attached to each cord informs the operator when each party replaces the receiver. These are called *Clearing-Out* or *Supervisory Signals.*

Thus the pair of cords and plugs with the associated ringing and listening keys and signals complete a cord equipment of which from 10 to 17 are placed in front of each operator on an appropriate shelf. Each operator is supplied with special lines extending to the chief operator, and to the trunk operators in other offices in exchanges of several central offices. These lines are called *Call Circuits* or *Order Wires*, and the switches connecting them to the operators' talking circuit, *Order* or *Call Wire Keys*. In the modern boards, a *Pilot Signal* placed in parallel with all the line and supervisory signals, calls the operator's attention, night or day, to the fact that business is to be transacted.

" *B* " *Positions.* — " B " positions differ from " A " positions in two important respects. There are no outgoing trunk jacks, as " B " operators receive calls from other offices only. Each trunk from other offices, equipped with a cord, plug, and clearing-out signal, terminates on a shelf before the " B " operator. Hence the " B " operator has only single cords and plugs, of which there are usually 25 or 30 per operator; a key to ring the called subscriber that is frequently arranged to be automatically tripped by the removal of the called subscriber's receiver from the hook, and no listening key.

2. DISTRIBUTING BOARD, OR CROSS-CONNECTING BOARD.

The distributing boards are frames, usually of iron, so arranged that all the wires of the wire plant may end on connectors (see Connector), on one side, and the cables from the switchboard on the other side, then by means of flexible pieces of wire called jumpers (see Jumper) it is possible at any time to change the connection of any wire plant wire to any switchboard wire without disturbing either the wire plant or the switchboard. There are usually two distributing boards, one called the Main Distributing Board, at which all the wire plant wires end, and to which extend all the switchboard cables. If a subscriber moves from one location to another, and must be connected to the office by an entirely different wire route, his switchboard number may be retained by changing the jumper connecting his former switchboard wire to the new wire plant wire. The main distributing board often carries the protection. The other distributing board, called the Intermediate Board, has no protection, and is frequently smaller, about two-thirds the size of the main board. From the main board the switchboard cables pass to one side of the intermediate board, and each wire is then exposed on a connector, then the same cables pass to the multiple in the switchboard. From the *answering jacks and signals* cables go to the other side of the intermediate board, thus by a jumper on the intermediate board any answering jack may be associated with any multiple jack. By this means busy subscribers may be distributed among different operators without changing their numbers or disturbing the multiple cables.

Running Board. — An inferior form of distributing board usually made by placing connectors on wooden strips often located at the bottom of and inside the switchboard frame.

3. POWER PLANT.

The entire apparatus installed in a central office to manufacture or transform electricity, and deliver it to the telephone switchboard. Power plants are of two types. (a) Those that take electricity from some commercial source, and transform the same into the desired electrical supply for telephone service, and (b) those that include a prime mover of some kind, and a dynamo driven therefrom. The first type comprises either a direct or alternating-current rotary transformer, usually installed in duplicate, for converting the electricity supplied by the commercial mains into the desired voltage and current for charging a storage battery. A ringing generator, also in duplicate, consisting of a dynamotor operated from the commercial mains and delivering alternating current for ringing at from 100 to 125 volts and about 16 cycles per second, of a size proportioned to the office, usually $\frac{1}{4}$, $\frac{1}{2}$, or 1 kw. A storage battery, usually of 10 to 12 cells in series (20 to 24 volts), sometimes of 20 cells (40 volts), of such capacity as to be able to carry the office for at least 36 hours without charging, on the basis of 10 ampere-hours per hundred originating calls. A switchboard equipped with the necessary switches to receive the commercial supply, and handle the transforming machinery and battery, together with all necessary instruments for measuring electricity, and the protection for the power circuits. In the second type, a gas engine, small water motor, or other prime mover is installed to supply power, a low voltage dynamo replaces the rotary transformers, and the ringing generators are operated from the battery. Both these forms of power plant include a fuse board that distributes and carries the necessary protection to guard the leads that supply electricity to the telephone switchboard, and a wire chief's desk supplied with all apparatus

necessary to test for, discover, and clear, any trouble that
may arise in any part of the telephone plant.

4. OPERATORS' FURNITURE.

At every central office proper quarters must be provided for the comfort of the operators. These include
necessary toilet facilities and bathroom, a dining room
equipped with gas stoves for heating tea and coffee, tables
and full set of table utensils and fittings for operators'
lunch, a locker room and lockers for the clothing of each
operator on duty at once, a parlor reasonably furnished
for the comfort of operators when resting, and a hospital
in which cases of illness may receive attention. This
may be a small ante-room provided with a sofa or cot, or
even a portion of the parlor partitioned off with a folding
screen. At the switchboard each operator must be provided with an adjustable chair. The manager must be
provided with an office properly furnished where he can
transact business with the public separate from the operating room.

The chief operator and monitors must be supplied with
chairs and desks equipped with telephone sets, signals,
and circuits to reach any of the operators at the board, the
wire chief and each other, and exchange lines for public
business. A card catalogue of subscribers, and a card
trouble record complete the list.

Branch Terminal or Bridging Board. — In the early
multiple switchboards each of the multiple jacks con-
tained a spring contact, and the jacks were arranged in
series, so that there were as many contacts as there were

jacks, any one of which could be opened by an infinitesimal particle of dirt. By setting the jack springs one on each side of the line, or in parallel, all the spring contacts were avoided, and the troubles due to open lines vanished. Boards with jacks so arranged are called *branch terminal* or *bridging boards*.

Busy Test. — Such an arrangement of a subscriber's circuit as will inform any operator at any position of the switchboard whether the line is being used. Often the busy test is accomplished by putting a battery on the sleeve of the plugs; then when a plug is placed in any jack, the rings of all other jacks of the same line are electrified, and any operator touching a charged ring with the tip of a connecting plug will hear a sharp "click" in her telephone.

Cable Forms. — Pieces of multiple wire cable that are to be attached to spring jacks, connectors, and the like, must have the ends of the various wires arranged in definite positions to match the apparatus to which attachment is to be made. When the wire ends are thus arranged, lashed, varnished, tested out, and marked, the cable is said to be *formed*.

Cable Run. — (a) Supports, preferably of iron, that carry and inclose the various aggregation of cables that connect the different pieces of apparatus in a central office. (b) The passage way or entrance or space including necessary supports provided to allow the cables of the wire plant access to the main distributing board.

Chief Operator. — The attendant who has general charge of a central office.

Circuit. — That combination of electrical apparatus

which permits of two parties to be placed within talking relations with each other. In the switchboard, circuits are known as "one-wire," "two-wire," or "three-wire" circuits, depending on the number of main conductors that are required for each subscriber's circuit through the jacks. Circuits are divided into (a) *Subscribers' circuits,* extending from the subscriber's premises and terminating in the switchboard at the exchange, including the answering jack and calling signal. (b) *Operators' Circuits* or *Cord Circuits,* or those which are used by the operator in the prosecution of the business of the exchange. (c) *Trunk circuits,* or those which are used to extend between two switchboards in different offices.

Cord. — A flexible electrical conductor attached to a plug.

Pair of Cords. — Each plug is supplied with a cord. When the two cords are united by connecting clips a flexible conductor is formed capable of instantly electrically uniting any two jacks. Cords are, therefore, usually used in pairs. Cords and plugs are further known as one, two and three-way, depending on the number of conductors they carry.

Cord Shelf. — A projection on the switchboard on which are placed the plugs and keys of the cord circuits, and from which the cords are suspended.

Color Scheme. — Switchboard cables are made up of single wires, wires in pairs, triples, or quadruples, depending on the nature of the circuit for which the cable is to be used. By dyeing the insulating material covering the various wires with different colors, according to a predetermined code, it is possible to identify at each end of a piece of cable the several wires without resorting to the laborious process of testing. Such a method of colored insulation is called a " *color scheme.*"

Connector. — Sometimes called " terminal," " line terminal," " clip," or " punching "; a U-shaped piece of thin metal having two ears, to which wires may be soldered, used to connect different wires or cables together. For distributing boards connectors are made up on maple blocks, each carrying 10 to 20 pairs.

Jack or Spring Jack. — A contrivance consisting of one or more springs and a ring held in an insulating frame, so designed that when a properly formed metal plug with a corresponding number of conductors is inserted into the ring, it may make contact with the various conductors formed by the ring and springs, and extend the circuits, of which they form a part, to corresponding conductors in a flexible cord attached to the plug. Jacks are divided according to their construction, into *one, two, three, four, five,* or *six-point* jacks, depending on the number of contacts, or conductors they contain, and according to the use to which they are put into *answering jacks, multiple jacks,* and *trunk jacks.*

Answering Jack. — The jack which is associated with the subscribers' calling signal, and which in multiple and divided boards is exclusively used by the operator in answering subscribers' calls.

Multiple Jacks. — Those which have no signal, and of which several are attached to each line, so distributed along the switchboard that each operator can reach every line. In multiple boards, multiple jacks are divided into *subscribers'* multiple jacks and *out-trunk* multiple jacks, depending on whether they are attached to subscribers' lines or to outgoing trunk lines. The subscribers' multiple is further divided into the " A " multiple, or those jacks which appear before " A " operators, and the " B " multiple, those which are placed in front of " B " operators.

Trunk Jacks. — Jacks which form the termini of trunk lines.

Jumper. — A piece of flexible wire, or metal strip, used as a usually more or less temporary connection between two parts of a circuit.

Key. — A switching device whereby different circuits may be connected or disconnected at pleasure.

Ringing Keys. — Keys used for placing ringing current upon lines.

Listening Keys. — Keys used to enable the operator to listen to subscribers' orders.

Combination Keys. — Keys which unite both above functions in a single piece of apparatus.

Order Wire or Call Circuit Keys. — Keys used by the operator to connect herself with a call wire.

" Listening in " or " Supervising." — The act of an operator in connecting her telephone with the circuit of two subscribers presumably talking together for the purpose of ascertaining whether the subscribers are conversing.

Operator's Load. — The amount of work, or number of messages, an operator can handle in a day's work. On modern switchboards an " A " position should handle from 1600 to 2000 messages per day. A " B " position from 2400 to 3000 messages. On a magneto board a fair " A " position load is from 1100 to 1300 messages, and for a " B " position from 1900 to 2100.

Manager. — The person in charge of a central office.

Peg Count. — The operation of counting the number of originating calls received in a central office during any consecutive 24 hours, for the purpose of ascertaining the load on the office. The result of a peg count when plotted with hours as abscissæ and number of originating messages as ordinates gives the "load" of the office.

"Position." — The space at a switchboard occupied by one operator, is also used to designate the entire collection of apparatus provided for an operator. Positions are divided into "A" and "B" positions.

Protection. — The apparatus provided to avert injury by abnormal currents to either (a) central office equipment, (b) sub-stations, or (c) wire plant wires.

(a) CENTRAL STATION PROTECTION.

Central station protection usually consists of a spark-gap and heat coil. The spark-gap is composed of two plates, preferably of carbon, one of which is grounded while the other is connected to the line wire, and separated about 1-200 of an inch, usually by a disk of perforated mica or silk. In a recess in one of the carbon plates a small lump of fusible metal (160 degs. solder) is often placed. This lump melts under the heat of the discharge, and flowing across the gap dead grounds the line. The heat coil consists of a small spool of resistance wire (5 to 50 ohms), in the center of which a movable pin is arranged to impinge on a ground plate. Normally, by means of fusible solder, the pin is held off the ground, but in case of an abnormal current of too low-potential to leap the spark-gap, the solder melts and drops the pin to the ground. Common battery boards must have the protection arranged to *open* the switchboard side of the line while *grounding* the wire plant side. Sometimes a light fuse (maxstat 1 ampere) replaces the heat coil, but is not so reliable. The entire protection for each line is placed on the line side of the main distributing board, the carbons and heat coils being mounted in springs to be readily replaceable after performing their functions.

(b) SUB-STATION PROTECTION.

The sub-station protection includes the same features as that for the central office, to which a fuse (7 amperes) is frequently added, the whole apparatus placed on a porcelain block located near the sub-station set (see Sub-station Set), or preferably on the wall of the building near the point where the line enters.

(c) CABLE PROTECTION.

At the point where a cable runs into an open wire line it is usual to place a fuse (3 to 7 amperes) between each cable wire and the open wire. When a cross takes place between a telephone wire and a wire carrying electricity at relatively high potential, the protection at the sub-station or central office operates and grounds the line, a rush of current then blows the fuse at the sub-station or at the junction of the cable or both, isolating the open wire from both the cable and the sub-station. Much difference of expert opinion exists as to the extent to which protection should be employed. There are many advocates of the entire omission of protection from underground lines and wide disagreement as to how far aerial lines should be guarded. Saving in expense is the only argument against the complete system as outlined. While protection devices are not infallible, to omit any of them is courting a risk that a conservative policy would shun.

Plug. — A small rod of brass shaped to fit a jack and used to connect the jack to a flexible cord. Plugs are divided into (a) answering plugs, or those which are only inserted in the jacks of subscribers making calls, placed next to the face of the switchboard on the cord shelf. (b) Connecting plugs, or those which are only inserted in the jacks of subscribers that are to be called, placed nearest the operator on the cord shelf.

Relay. — An electromagnetic device consisting of an electromagnet that when excited attracts a movable armature, thus completing a circuit. Relays are divided into (a) *Line relays*, or those which are used by subscribers for signaling the exchange ; (b) *Cut-off* or *Controlling relays*, or those which are used to isolate a subscriber's line from its answering jack and signal ; (c) *Supervisory relays*, or those which control the supervisory signals. (d) *Pilot relays*, or those common to a number of signals acting as a general guard in case of failure of any individual relay.

Signals. — A device for calling the attention of an operator. Signals are divided into *Line* or *Calling* signals, or those attached to subscribers' lines for the purpose of notifying an "A" operator that a subscriber desires a connection. *Clearing out* or *Supervisory* signals that are attached to the cord circuits of both " A " and " B " operators, chiefly used to notify operators that conversation is completed; and *Pilot* signals, or those which are common to, or are in parallel with, a group of other signals, and serve to call attention to the fact that some other signal is displayed. Example: A night bell connected to all the signals of a magneto board. Signals may also be classed by their operation into *Manual* signals, or those that after being displayed must be restored by hand to their normal position, and *Automatic* or *Self-restoring* signals that take care of themselves. The ordinary annunciator, an electromagnet that when excited attracts its armature, and allows a shutter to fall by gravity, displaying a number, is an example of the first class. Such a signal is usually called a " *Drop.*" Of the second class the miniature electric lamps of the modern switchboards form the best example.

Supervisor. — An operator who has charge of a division of the switchboards, supervising a group of operators, and reporting to the chief operator.

Terminals. — Telephone circuits usually contain many different pieces of apparatus, and frequently several kinds of cable or wire. When one piece of apparatus or kind of wire is to be joined to another, a great variety of connecting devices are used, all of which pass under the generic name of *terminal.*

Wire Chief. — The attendant who has charge of the power plant, and supervision of the switchboard. In many cases the wire chief has also more or less charge of the maintenance of the aerial wire plant, and sometimes is expected to run lines for new subscribers.

(d) THE WIRE PLANT.

Structurally the wire plant may be divided into (a) aerial lines, (b) underground lines.

(a) Aerial lines comprise all of the wire plant that is above the ground.

Aerial lines may be divided into (1) open wire lines and (2) aerial cables.

(1) *Open Wire Lines* are those in which individual wires are supported upon insulators set on cross arms carried upon line poles.

(2) *Aerial Cables* consist of a number of insulated conductors bunched together in a lead sheath and supported upon cross arms upon a pole line. Aerial cables are usually 10, 25, 50, and occasionally 100 pair.

Anchor Pole. — The last pole at the end of an aerial line upon which the wires are terminated, and which must be sufficiently strong to sustain the longitudinal stress of the wires.

Balcony. — A small platform built upon a pole carrying a cable head upon which workmen may stand, for the purpose of access to the cable head.

Common Return (sometimes called *McClure System*). — It was formerly customary to use the earth as one conductor for every telephone line, but inductive disturbances became so severe as to compel the resort to a complete metallic circuit to afford the best service. The use of one wire of somewhat larger section than the ordinary line wire serving as common conductor for a number of lines (a partially successful compromise between the more expensive complete metallic circuit and defective ground) has received the preceding title.

Distributing Pole. — A pole from which a number of drop wires pass to sub-stations.

Messenger, sometimes called " *strand.*" A wire rope carried upon a special cross arms (usually angle iron) of a pole line and used for supporting an aerial cable.

Transposition. — By arranging the wires of an open-wire line so that instead of proceeding straight along on the same pin through the entire line, each wire frequently changes from one pin to another in such a manner as to thoroughly interlace all the circuits among themselves, inductive disturbances may be prevented. Such interlacing is called " *transposition.*"

Terminal Pole. — A pole carrying a cable head or terminal, sometimes called a distributing pole.

(b) UNDERGROUND PLANT.

The underground plant comprises all of the conductor installation which is placed below the street level.

Block System. — The plan or design of a telephone cable plant involving the extension from the central office of one or more main cables that are subsequently split into a number of subordinate branches and extended from one distributing point to another in series.

Cable Entrance or *Cable Run.* — Where cables, either aerial or underground, leave the street and enter the central office building, proper facilities must be provided to convey them from the street to the cable side of the main distributing board. All the space taken for this purpose, together with all the necessary supports or fittings to hold the cable, is called the *cable entrance,*

Cable-head, " *Terminal*," or " *Cable Box*." — A device into which a cable is extended and to which the lead sheath may be soldered, forming a water-tight joint. The cable head is usually in the form of a long rectangular box, preferably of iron, upon the inside of which the cable wires may be fanned out and attached to binding posts, which project through the sides of the box, the whole being supplied with a water-tight cover. By this means the individual pairs of an aerial cable may be connected to the open wire lines by means of short pieces of wire called " bridles " or " jumpers."

Distributing Box. — A device similar to a cable head, through which a main cable passes, which permits a portion of the cable pairs to be taken off and distributed to adjacent sub-stations. Distributing boxes are now becoming obsolete, and are replaced by " Y " splices.

Lateral Cable (sometimes called "Subsidiary Cable").— That portion of the underground cable plant which extends from a main cable or distributing box to a terminal pole or building terminal.

Conduit. — A structure installed beneath the street level for the purpose of carrying electrical cables. Conduits may be divided into (a) drawing-in systems, so planned as to permit of introducing or removing the cable at any time after the structure is completed; (b) built-in systems, in which the cable is placed at the time the conduit is constructed and cannot thereafter be removed or changed. Built-in systems are gradually becoming obsolete.

Duct. — Such a longitudinal portion of the conduit as is intended for the reception of one or more cables. The tendency of present practice is to so construct conduits that each duct shall receive only one cable. Ducts are divided into:

(a) Main ducts which carry feeder cables or main cables, and (b) lateral ducts, or those which branch from the main conduit to a distributing pole or office terminal.

Encasement. — The concrete or other material with which the ducts forming a conduit are surrounded for the sake of mechanical strength and protection.

Fish Wire. — A wire which is placed in a duct for the purpose of drawing in a cable.

Main Cables (sometimes called " feeder cables "). — That portion of the underground cable plant extending from the office, along a highway to a distributing box or lateral cable.

Manhole (sometimes called " vault"). — A chamber constructed beneath the street level, and giving access to a portion of the conduit for the purpose of introducing and removing cables.

Office Manhole. — A manhole or vault, usually much larger than standard size, placed as close as possible to the central office to which all the conduit system converges, and from which the cable entrance extends into the office.

Pot Head. — A peculiar kind of cable terminal, consisting of a piece of lead pipe somewhat larger than the cable sheath, to which it is soldered, making an air-tight joint. Into this sleeve the cable conductors are extended and connected to okonite or other water-proof wire. The sleeve is then filled with a water-proof compound melted and poured in, thus effectually sealing the cable from the incursion of moisture, yet permitting the conductors to be extended at pleasure.

Rodding. — The operation of introducing into a duct a fish wire to facilitate the future introduction of a cable.

Section. — That portion of a conduit extending between adjacent manholes.

Underground Cable. — A collection of conductors surrounded with a lead sheath and designed for service in underground conduits. Underground cables are usually 25, 50, 100, 120, 150, 200, and 300 pair.

"Y" Splice. — The separation of a feeder cable into two or more parts by splicing to the cable one or more branch cables. This splice is made by soldering to the main cable sheath lead sleeves to which the smaller cables are attached, thus forming a " Y." A cable thus divided is said to be *tapped*, and each part is often termed a *tap*.

Battery Wires (also called " power wires "). — Lines which are used for the purpose of carrying battery currents or electricity for other purposes than talking.

Branch Exchange Line or Trunk. — A line extending from a central office to a branch exchange.

Call Wires (also called " order wires "). — Circuits used to permit operators to talk to each other for the purpose of giving instructions regarding exchange business.

Combination Line. — A line which is used both for talking and for telegraphing.

Distributing Line. — A line branching from a main lead running to various sub-stations. Distributing lines are usually aerial. As in many cases such lines are built in alleys they are frequently termed " alley " lines.

Drop Wires. — The portion of the subscribers' line

which extends from the outside of the wall of his premises to the nearest wire run.

Long-Distance Lines. — Toll lines which extend over such a distance as to pass through the offices of a number of companies.

Main Line (sometimes called a " through line," a " trunk line," or "main lead"). — A portion of the wire plant which extends from the exchange to the distributing lines.

Private Line. — A line extending between subscribers not used for exchange service.

Subscribers' Line. — A wire or pair of wires used to give service to subscribers. Sometimes called an exchange line.

Toll Lines. — Trunk lines extending outside the territory of an exchange, and upon which an extra rate or charge for service is made.

Trunk Line. — A wire or pair of wires extending between two central offices used to transmit originating calls in one office to subscribers in the other.

(e) SUB-STATION.

Subscribers' Instrument. — This term includes the complete talking and signaling apparatus used by the subscriber, and comprises a hand telephone, receiver, call bell and means of signaling. Subscribers' sets are commonly classified into cabinet desk sets, or instruments which are mounted in a kind of desk or cabinet, at which one using the telephone may conveniently sit and write; desk sets, or those which may be semi-portable by the use of a flexible cord, and so may be placed upon a table or desk;

extension sets, or those consisting of a duplicate set (often of a different pattern), placed on a line that is an extension or addition to the first set, usually running to an adjoining room. The circuit in such instruments is frequently so designed that the first instrument upon the line may receive the call, and the extension instrument subsequently used on a secondary signal. The use of such circuits is common in offices where a clerk or other employe first receives the calling signal and subsequently summons his principal. Party-line sets, or instruments which are designed to be operated upon lines having more than one subscriber; wall sets, or those which are supported by being fastened to a partition wall. Each set consists of the following parts:

Transmitter, or apparatus used by the subscriber in talking to another party.

Receiver (also called " *Telephone* "). — The instrument for receiving conversation. When held in the hand receivers are called " *hand telephones*." When supported at the ear by a band or cap on the head, receivers are called " *head telephones*." Receivers are divided into common battery receivers and magneto receivers. Common battery receivers are of lower resistance than magneto receivers.

Ringer or *Bell*. — The device for making an audible signal at the sub-station, usually consisting of an electromagnet carrying a pivoted armature, to which a bellclapper is attached, the passage of an alternating current through the magnets, causing the armature to vibrate, thus sounding the bell. Ringers may be high wound, 200 to 1600 ohms (for bridging bells), or low wound, 25 to 150 ohms (for series bells).

Generator (sometimes called "magneto-generator" or "magneto"). — A small hand-operated alternating-current dynamo capable of giving from 10 to 20 cycles per second, usually at a potential of from 50 to 125 volts, used for sending alternating currents to operate ringers. Generators may be high or low wound, depending upon the service to which they are to be devoted, high-wound generators giving a greater e. m. f. than the low wound.

Back Board. — Strictly the wooden board to which subscribers' sets are attached, but frequently expanded to mean the entire woodwork that is used for supporting a sub-station outfit.

Bell or *Call Bell.* — Usually understood to mean the entire signal bell, including the gongs, clapper and ringer magnets. Bells are usually designated by the resistance to which they are wound, as 40, 60, 80, 100, or 1000 bells.

Biased Bell. — This term is sometimes applied to a polarized bell, but strictly relates only to bells in which the ringing armature is pulled in one direction by a spring, and thus given a set in one direction. Pulsating currents of one polarity will affect such a bell, while it is unaffected by similar pulsating currents of opposite polarity. Used as a selective signal for party lines.

Hook or *Telephone Hook.* — A pronged support arranged on a sub-station set to hold the receiver. The hook also forms an automatic switch operating contacts in the sub-station circuits.

Induction Coil. — A coil consisting of two electrically separate windings forming a small step-up transformer. The low winding of the induction coil is placed in circuit with the transmitter, while the high winding is placed in series with the line. The office of the induction coil is to raise the potential of the impulses set up by the transmitter.

Polarized Bell. — A bell in which the armature or ringer cores or both are permanently magnetized so that the armature is held in a predetermined position with reference to the ringing magnets. Currents of one polarity will then cause such bells to ring, while currents of the opposite polarity will not affect it. Used in connection with a party line as a selective signal.

Wall Plugs. — Wooden plugs placed in holes drilled in house walls to which sub-station instruments are fastened, and by which they are supported.

CHAPTER III.

THE CENTRAL OFFICE SITE.

SLIGHT allusion has been made to the impossible complexity of a telephone plant devoid of a central office, but as even some telephonists rarely recognize the full signifi-

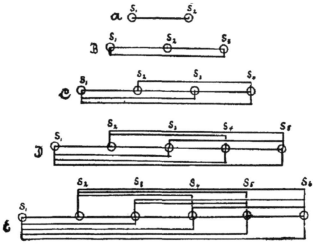

Fig. 1.—Diagram of Circuits Between Sub-stations.

cance of "Central," and as the general reader seldom thinks of it save to anathematize some unfortunate operator, it seems desirable to point out in detail the enormous economic value of this factor.

Consider Fig. 1, in which at A, B, C, D, and E, five separate groups of sub-stations are shown as indicated by small circles. At *A* there are two subscribers, at *B* there are three subscribers, at *C* four, at *D* five, and at *E* six subscribers. Imagine in each case that between the subscribers, taken in pairs, a wire is strung in such a manner as to give the least number of lines to enable each one to talk to all the others in both directions. Such wires are represented by the full black lines joining the circles. At *A* a single one joining 1 and 2 will suffice. At *B* there must be one line between 1 and 2, and one between 2 and 3, and one from 1 to 3. At *C* there must be one between 1 and 2, and one from 2 to 3, one between 3 and 4; also one each from 1 to 3, and from 1 to 4, and lastly one from 2 to 4. Continuing this reasoning it is easy to see that if there are *m* subscribers in any territory, and supposing them to be numbered from 1 to *m*, consecutively, then there will be $m - 1$ lines from the first subscriber; $m - 2$ from the second; $m - 3$ from the third, etc., to one line between the $m - 1$th subscriber and the *m*th subscriber. The total number of lines will evidently be the sum of all these, or

$$\Sigma m - 1 + m - 2 + m + 3 + \text{etc.} + 1.$$

This is the sum of an arithmetical progression, of which the first term is $m - 1$, the last term 1. and the common difference 1. The familiar formula for the sum of a series is

$$S = \frac{n}{2} (a + e), \qquad \text{(Eq.}$$

in which *S* is the desired sum; *n* the number of terms; *e* the last term, and *a* the first term. In this case

$$n = m - 1; \quad a = m - 1; \quad e = 1.$$

Therefore,

$$S = \frac{m-1}{2}(m-1+1) = \frac{m^2 - m}{2}. \quad \text{(Eq. 2)}$$

In Table No. 1 the value of S is calculated for a few of the most commonly occurring groups of subscribers up to 5000, to illustrate the rapid increase in complexity that such a system would cause.

TABLE No. I.

Showing relation between number of subscribers and number of lines needed with no central office.

NUMBER OF SUBSCRIBERS.	NUMBER OF LINES.
2	1
10	45
100	4,950
500	124,750
1,000	499,500
5,000	12,497,500

Cases where there are a thousand subscribers or more grouped together are numerous, needing, as shown by Table No. 1, nearly 500,000 lines. A 200-pair cable, the largest now commonly used, needs (including space for ducts) about 25 square inches, so the 500,000 lines, should they pass a single point, would need 2500 200-pair cables, occupying 435 sq. ft. space in the street, taking in its most compact form, at least 20 ft. square. Even the astonishing maze of pipes and other underground structures of New York streets would fade into utter insignificance beside such a scheme. Nor is the resort to an

aerial line conceivable. A line with ten cross arms carrying ten wires each is considered the largest justified by
good practice, hence 5000-pole lines would be needed for
a petty exchange of a thousand subscribers. By similar
reasoning, it is easy to see that the first subscriber must
have 999 lines, requiring five 200-pair cables or ten-pole
lines to care for his wire plant only. The perplexities
are not yet exhausted. Supposing one new subscriber to
be added to an exchange of a thousand, then a thousand
new lines must be installed.

A similar huge plant outlay, and corresponding space
at each sub-station, would be needed to enable connections
to be made between subscribers. As subscriber No. 1 has
999 lines, he must have the same number of terminals so
placed that any one can be connected to each of the
others. At present, the jack with its connecting cord and
plug is the most compact known form of switching apparatus, but a thousand jacks with their companion appliances would require a switchboard as large as a folding
bed, costing at least $3,000. To be sure, but one pair of
cords and plugs would be needed, and no supervisory
signals, but all other switchboard apparatus would be essential. On each subscriber would fall the burden of doing
all the switch work necessary to carry on his telephonic
correspondence, no light task for people telephonically
busy, and then the switchboard and wire plant must be
maintained and kept in good working condition, a matter
needing expert knowledge and unremitting attention.
So, from all standpoints, "Central" is shown to be a
great economic factor in telephony, saving not only a vast
investment over any other conceivable method of attaining

the same result, but relieving the subscriber of an other-wise impossible quantity of irksome labor, accomplishing the desired end, with, on the whole, wonderful speed, accuracy, and economy, facts which it is well to bear in mind the next time the gentle " Hello Girl" is a second or two longer in getting a desired party, gives, perchance, a wrong number, or cuts off in the midst of an important sentence.

If, then, the central office is admitted to be a necessity, the best situation therefor is an important consideration. Unlike other location problems in electrical engineering, there are no questions of fuel or water supply, and no attention need be devoted to electrical losses in the radiating lines, so consideration relieved of these details may be focused sharply on the pure cost of the necessary installation, including therein the factors of real estate (land and buildings) and electrical plant.

If the operating company buys real estate (land and buildings), or if it purchases land and builds, it must treat itself as a landlord, and charge against operating expense a sufficient annual sum to cover real estate maintenance, namely, interest on investment in land, interest and depreciation on buildings, expense of janitor service, light, heat, general building maintenance, etc., for the same. If the company rents office quarters from other real estate owners, the annual rental therefor covers the same items, plus a probable profit to the owners, and must be charged against operating costs in the same manner. Thus, in so far as the actual annual operating expense is concerned, it matters little whether the company, or other parties, are the owners of the necessary real estate, except that if the company owns, it may save to itself the interest on the real estate investment, or any other expense in the nature of a profit on the real estate transaction, and charge against operating only such a sum as is actually needed to perpetually maintain the buildings in proper condition, leaving the general profits of the business to care for any desired profit on the real estate portion of the investment.

There are other and weightier reasons in favor of a telephone company owning its own offices. A central office of any magnitude is a large, complicated, and expensive installation. The nature of the wire plant renders a removal very difficult, costly, and exceedingly detrimental to the service, even if conducted with the greatest skill and care, so permanence in location is of vital importance. Ordinary buildings are ill adapted to the uses of a central office, and if adopted, it must be at considerable sacrifice of economy in plant installation cost, or at an increase in operating expense. Further, telephone apparatus is peculiarly sensitive to injury by fire, as the most insignificant one may completely paralyze the service for a considerable period of time, so particular pains should be taken to protect therefrom. Central offices, therefore, are preferably located in buildings more than ordinarily fireproof, thus even the presence of other occupants courts a fire risk that it is well to avoid. It is, therefore, safe to assume that ownership by the telephone company of such real estate as shall be adequate for at least the more important offices, say, those of 800 lines or over, represents the best present practice.

The design of the central office, including therein the

choice of the necessary site, will affect the cost of four
items in the telephone plant, namely, the investment in
real estate, in buildings, in switchboard, and in wire plant.
So far as a discussion of the best location is concerned,
two of these, namely, the type of switchboard and building,
will remain essentially constant. For all kinds of switchboard need about the same space and other requirements,
and the cost of buildings at any given time will be essentially the same, whether placed on one site or another of a
particular city, provided, of course, that there are no exceptional peculiarities in the locations considered. But it
is very evident that the cost of land will vary very widely
in the same town, sites in the business portion costing
many fold the price of equivalent areas in the outskirts.
It is also evident that there must be some point within
every collection of subscribers, such as would be presented
by the groups in every city or town, from which the expenditure of wire to reach all the subscribers will be a
minimum. To locate the office at this point, generally
called the " telephonic center," will certainly require the
least wire plant investment, and involve the smallest annual
charge for depreciation and maintenance, because both the
quantity of plant and investment therein is a minimum.
As the office is moved away from the telephonic center,
the investment in wire plant increases in a manner easy
to calculate, but the cost of real estate will also change in
some manner to be ascertained only by inquiry on the
ground. The problem presented to the telephone engineer is then this : Given the location of the subscribers
and cost of various central office sites, to determine, first,
the telephonic center ; second, the rate of increase in in-

vestment and annual expense of the wire plant as the office recedes from the telephonic center; and, third, to select such a site as will make the sum of the investment in land and wire plant, and the sum of the annual charges on both these items, a minimum.

CHAPTER IV.

DETERMINATION OF THE TELEPHONIC CENTER.

THEORETICALLY, it is easy to imagine that the various
subscribers in any group could be connected to the office
by air lines, and under this supposition the problem of
determining the telephonic center is that of ascertaining
the position of such a point, within a group of given
points, as will make the sum of all the straight lines
drawn from the desired point to each of the given points
a minimum. If this sum be assumed to be wire mileage,
it is impossible to further reduce the quantity of wire
needed to reach all subscribers. The mathematical solution of this problem is possible for a few points, but when
a considerable number come under consideration the
necessary equations become too complicated for solution.
Under all usual circumstances, it is impracticable to
design subscribers' wire routes so that they shall fall
along air lines drawn between the telephonic center and
the premises of each of the various subscribers. On the
contrary, in all towns and cities, it is necessary to follow
the general highways (streets or alleys), for it is impracticable to obtain any considerable right of way in other
locations. In most cases, particularly in all of the more
modern cities, the majority of streets, alleys, and other

highways, are laid out, approximately, rectangularly with reference to each other, at least, so far as pertains to such a group of subscribers as is usually placed in a single office. For the practical solution of the problem, it is, therefore, sufficiently safe to assume that the wire routes *will* follow existing streets, and that these *are* rectangular to each other, so the problem is reduced to the location of such a point within a group of given points as will make the sum of the distances from the desired point to all the given points, measured parallel to a pre-determined set of rectangular co-ordinate axes, a minimum. Mathematically, it might be possible to select some other direction for the axes which would further minimize the necessary wire mileage; but, as it is impossible to change the arrangement of the streets, and impossible to obtain other rights of way, the only feasible solution of the problem lies in the assumption of such axes as are indicated by the direction of the streets, and it is unnecessary to consider the saving in wire mileage that could theoretically be made by the consideration of other axes.

Diagonal streets are of occasional occurrence, particularly in the older towns. The possible availability of such diagonal streets must not be lost sight of, and must be considered, as far as it is cheaper by their employment to reduce wire mileage. The solution of the problem will, however, be simplified by at first considering the determination of the telephonic center in a case with purely rectangular streets, and then afterwards ascertaining the *effect* which the introduction of one or more diagonal routes would have upon the previously ascertained wire mileage.

Let Fig. 2 represent the plane of any city, and let s_a, s_b, s_c . . . . s_n represent the number and location of the various subscribers. The subscripts a, b, c . . . . n, denote the number of subscribers at the locations a, b, c . . . . n, and may be varied in value in each from 1 to n, at pleasure. The distribution of s_a, s_b, s_c . . s_n over the plane may

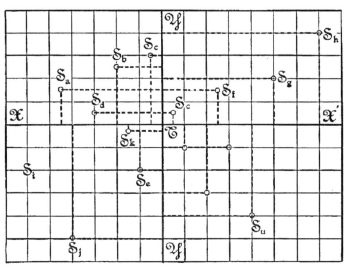

Fig. 2. — Determination of Telephone Center.

be in any manner whatsoever. Let T be the desired point from which the sum of all lines to s_a, s_b, s_c . . . . s_n shall be a minimum, then T will, by definition, be the telephonic center. Let the lines $X\,T\,X^1$ and $Y\,T\,Y^1$, drawn rectangularly through T, indicate the prevailing direction of the streets or highways that are available for wire routes, so that $X\,T\,X^1$ and $Y\,T\,Y^1$ become the co-ordinate axes with origin at T, parallel to which the distances of all subscribers shall be measured; then, by hypothesis, T must be such a point as will make the sum of all the distances to all subscribers a minimum.

Symbolically, this condition is expressed as follows:

$$s_a(x_a+y_a)+s_b(x_b+y_b)+ \cdots \cdots s_n(x_n+y_n)=\text{a minimum, (Eq. 3)}$$

in which

$$x_a, x_b \ldots x_n \quad \text{and} \quad y_a, y_b \ldots y_n$$

are the co-ordinate distances from the point T to each of the points

$$s_a, s_b \ldots s_n.$$

Also, let

$$s_a + s_b + \cdots \cdots s_n = U + V, \qquad \text{(Eq. 4)}$$

in which U and V are, respectively, the number of subscribers on one side and on the other side of either of the co-ordinate axes (taken singly) through the desired point. If there be $a, b \ldots \ldots n$ subscribers at the points $s_a, s_b \ldots \ldots s_n$, it is legitimate to replace each term in Equation 3 by the corresponding x's and y's repeated as many times as there are units in $a, b \ldots \ldots n$. Or, Equation 3 may be written

$$x_a + x_a + x_a + \cdots \cdots a \text{ times} + x_b + x_b + x_b + \cdots \cdots b \text{ times}$$
$$+ x_c + x_c + x_c + \cdots \cdots c \text{ times} + x_n + x_n + x_n + \cdots \cdots n \text{ times}$$
$$+ y_a + y_a + y_a + \cdots \cdots a \text{ times} + y_b + y_b + y_b + \cdots \cdots b \text{ times}$$
$$+ y_c + y_c + y_c + \cdots \cdots c \text{ times} + y_n + y_n + y_n + \cdots \cdots n \text{ times}$$
$$= \text{a minimum.} \qquad \text{(Eq. 5)}$$

The following theorem will aid in obtaining the solution of Equation 5:

Let
$$z = f(x, y), \qquad \text{(Eq. 6)}$$

in which x and y are independent variables as in the present problem, then

$\delta z = f(x + \delta x, y) - f(x, y) + f(x, y + \delta y - f(x, y).$ (Eq. 7*)

Equation 7 is the sum of the partial differentials with respect to both x and y.

The partial differential with respect to x is

$$f(x + dx, y) - f(x, y) = \frac{\delta z \, \delta x}{\delta x}, \qquad \text{(Eq. 8)}$$

and the partial differential with respect to y is

$$f(x, y \, \delta y) - f(x, y) = \frac{\delta z \, \delta y}{\delta y} \qquad \text{(Eq. 9)}$$

For brevity let

$$\frac{\delta z}{\delta x} = M \quad \text{and} \quad \frac{\delta z}{\delta y} = N,$$

then

$$\delta z = M \, \delta x + N \, \delta y. \qquad \text{(Eq. 10)}$$

The symbol $\frac{\delta z}{\delta x \, \delta y}$ indicates the complete differential of z with respect to both x and y. The symbol $\frac{\delta z}{\delta x}$ indicates the partial differential of z with respect to x, and the symbol $\frac{\delta z}{\delta y}$ the partial differential of z with respect to y.

Differentiate M with respect to y obtaining,

$$\frac{\delta M}{\delta y} = \frac{\delta^2 z}{\delta x \, \delta y}. \qquad \text{(Eq. 11)}$$

and differentiate N with respect to x obtaining,

$$\frac{\delta N}{\delta x} = \frac{\delta^2 z}{\delta x \, \delta y}. \qquad \text{(Eq. 12)}$$

If any expression, such as $f(x, y)$, be differentiated first with respect to x and y, and then with respect to y and x, it is evident that the results will be equal, for the results of the same process applied to the same function must be

---

* See Edwards's Calculus, pp. 112, 155.

the same, irrespective of the order in which the variables are taken, hence

$$\frac{\delta M}{\delta y} = \frac{dN}{\delta x}\,, \qquad\qquad \text{(Eq. 13)}$$

and this equation determines the condition which must exist in Eq. 10 in order that a differentiation shall be possible.*

To apply this theorem to the problem in hand suppose $f(x, y)$ in Eq. 6 to be the sum of the series of Eq. 5. This sum is the total wire mileage to all subscribers, which by definition must be a minimum. Let x and y (without subscripts) stand for the *average* ordinate and *average* abscissa; as U and V are the number of subscribers on each side of either axis, respectively, then $(U + V)(x + y)$ is the total wire mileage, and

$$z = f(x, y) = (U + V)(x + y). \qquad \text{(Eq. 14)}$$

This expression must now be differentiated, equated to zero, and solved after the usual method applicable to problems in maxima and minima. By equations 6 to 13 it has been shown that the differentiation is possible only when the condition expressed in Eq. 13 holds true; therefore, differentiate Eq. 14 by differentiating first with respect to X and then with respect to Y, and add the two so obtained partial differentials, thus:

$$dz = \frac{\delta(U + V)(x + y)\,\delta x}{\delta x} + \frac{\delta(U + V)(x + y)\,\delta y}{\delta y}. \quad \text{(Eq. 15)}$$

This expression corresponds to Eq. 10,

$$\frac{\delta(U + V)(x + y)}{\delta x} \text{ representing } M,$$

and $\qquad \dfrac{\delta(U + V)(x + y)}{\delta y}$ representing N.

---

* See Johnson's Differential Equations, p. 22.

Now, to determine the possibility of differentiation according to the condition of Eq. 13, differentiate M with respect to y, and N with respect to x, and test for equality, thus:

$$\frac{\delta}{\delta y}\left[\frac{\delta(U+V)(x+y)}{\delta x}\right] = \frac{\delta}{\delta x}\left[\frac{\delta(U+V)(x+y)}{\delta x}\right] \quad \text{(Eq. 16)}$$

Performing the operations indicated,

$$(U+V)\frac{\delta(x+y)}{\delta y\,\delta x} = (U+V)\frac{\delta(x+y)}{\delta x\,\delta y}. \quad \text{(Eq. 17)}$$

But,

$$\frac{\delta(x+y)}{\delta y\,\delta x} = \frac{\delta(x+y)}{\delta x\,\delta y};$$

therefore,

$$U+V = U+V, \quad \text{(Eq. 18)}$$

and it is proved that Eq. 15 *can* be differentiated. For the final solution the first differential coefficient must be placed equal to zero and the resulting equation solved, thus:

$$\frac{\delta(U+V)(x+y)}{\delta x} + \frac{\delta(U+V)(x+y)}{\delta y} = 0. \text{ (Eq. 19)}$$

Differentiating,

$$U+V+U+V = 0. \quad \text{(Eq. 20)}$$
$$2U+2V = 0. \quad \text{(Eq. 21)}$$
$$U = -V. \quad \text{(Eq. 22)}$$

There are many problems in maxima and minima, of which the solution indicates not a numerical conclusion, but a certain and very definite *condition* that must obtain in order to produce the desired result. Such is the case in the problem under consideration. U and V are the *number of subscribers* on each *side* of either one of the co-ordinate axes, and the expression $U = -V$ of Eq. 22

means *that if there are an equal number of subscribers on each side of one axis the sum of the perpendicular distances from that axis to all subscribers will be a minimum.* The same relation must obviously hold true with respect to another axis drawn at right angles to the first one. Therefore, Eq. 22 shows the *relation which the subscribers sustain to the co-ordinate axes when the sum of all their perpendicular distances to the axes is a minimum.*

In other words, if a line be drawn parallel to the prevailing direction of the streets, in such a position as to place numerically one-half the subscribers on one side of the line, and one-half on the other, the sum of the perpendicular distances from this line to all the subscribers will be less than the sum of the perpendicular distances from all the subscribers to any other line. Having thus located one axis, it is evident that the same condition will define the position of the other one to be drawn at right angles to the first, and as the telephone center must be on both axes simultaneously it must lie at the intersection of the two lines.

CHAPTER V.

ILLUSTRATIONS OF THE SITUATION OF THE TELEPHONIC CENTER.

THE demonstration of the preceding chapter gives rise to a very facile method for ascertaining the telephonic center for any given city. On any fairly good map, plot the location of existing or prospective subscribers. Find the total number thereof, then place a ruler parallel to one of the prevailing street directions, and move it over the map till it has crossed one-half the number of subscribers. A line drawn by the edge of this rule will be one axis. Place the rule at right angles to the first line and repeat the process, obtaining the second axis, with the telephonic center at the intersection of the two. To illustrate this method, refer to Fig. 3, in which the small black circles represent groups of subscribers. One subscriber is supposed to be placed at each of the points 1, 2, 3, 4, 5, 6, 7, and 8, but the reasoning is the same for any number of subscribers at these points. Assume the lines of the section paper to represent the prevailing direction of the streets. Then, by the preceding rule, the line Y T Y must be drawn parallel with the lines of the section paper, and in such a location as to place four subscribers on the right hand of it, and four on the left hand. Similarly, the line $X\,T\,X$ must be drawn at right angles to $Y\,T\,Y$, and at such a location as to place four subscribers above the line and four below it. Then T is the telephonic center, $Y\,T\,Y$ and $X\,T\,X$ are the axes, shown in full black line in the figure, and the sum of the perpendicular distances from $Y\,T\,Y$ and $X\,T\,X$ to 1, 2, 3, 4, 5, 6, 7, and 8 is less than the sum of the same distances to any

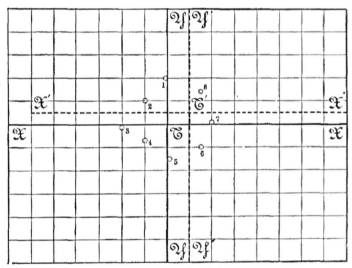

Fig. 3.— Determination of Telephonic Centers.

other pair of lines that can be drawn parallel to the section paper lines.

In this illustration, subscribers 1 and 5 and 3 and 7 are supposed to lie infinitely near the axes of Y and X, respectively, but on opposite sides of these lines, so that if either line were moved slightly in either direction it would pass across one subscriber. Now, suppose the axis Y to

shift parallel to itself one unit, say, to the right, then
the perpendicular distances to subscribers 1, 2, 3, 4, and
5 will be *increased* one unit each, and the distances
to 6, 7, and 8 will be *decreased* one unit each. There
will, therefore, be a total gross increase of five units,
and a decrease of three units, or a net increase of two
units to the total mileage of all subscribers from this axis.
Assume that the amount that the axis is shifted *decreases*

TABLE II. — COMPARISON OF WIRE MILEAGE FOR FIG. 3.

SUBSCRIBER'S NUMBER.	DISTANCE FROM			
	$Y T Y$	$X T X$	$Y^1 T^1 X^1$	$X^1 T^1 X^1$
1	0	19	10	14
2	9	9	19	4
3	19	0	29	5
4	9	6	19	11
5	0	14	10	19
6	14	9	4	14
7	19	0	9	5
8	14	13	4	8
Totals	81	70	104	80
	. . .	84	. . .	104
Total distance from axes,	154	154	. . .	184
	. . .	. . .	. . .	154
Increase	. . .	. . .	. . .	30

till it is equal to the smallest assignable quantity, then
the net increase in mileage will correspondingly decrease;
but it is evident that so long as the motion of the axis is
greater than zero the increment to the mileage will be a positive quantity greater than zero, hence the position $Y T Y$
gives a minimum mileage. The same reasoning may be applied to the position of $X T X$, thus proving that T is the
telephonic center. In Table 2 the perpendicular distances of

the subscribers from the axes, in terms of the ruling of the section paper as a unit, is given both for the location through the telephonic center and a new location at X^1 $T^1 X^1$ and $Y^1 T^1 Y^1$, shown in dotted lines. In this example the Y axis was shifted ten divisions of the section paper, and the X axis five. The resulting increase in mileage is 30 units; but if the motion of either axis is varied in any amount, it is easy to see that the mileage increment will be proportionally changed.

If there are an uneven number of subscribers in the territory the preceding rule of drawing the axes in such a manner that numerically one-half the subscribers lie on each side of each axis still holds true. For, in order to place an *equal* number of subscribers, one *each side* of *each axis*, each axis must pass *through* at least one subscriber. When a subscriber is located *on an axis*, the *distance* from that axis to the subscriber becomes zero, and the function of x or y, as the case may be, vanishes, and in so far as the axis under consideration is concerned the subscriber vanishes, and thus the condition of Eq. 22 holds true for those that remain, requiring one-half the *balance* to be placed on each side. To illustrate, turn to Fig. 4. Assume the plane of the paper to represent any territory, and the black circles numbered from 1 to 11, inclusive, to stand for the subscribers, one at each point, and that the section paper lines coincide with the general direction of streets. Enumerating the subscribers 11 are found. To locate the line $Y T Y$, so that there shall be an equal number of subscribers on each side of and parallel to the section lines, can only be done by drawing it to pass through No. 11. $X T X$ must be drawn perpendicularly to $Y\,T\,Y$, and so situated as to place half the subscribers above the line and half below it. This can only be done by drawing it through No. 3, and the telephonic center is found at the point T. In this illustration the ordinate of No. 11 is 20, using the squares of the section paper as units, and the abscissa o, and for No. 3 the ordinate is o and

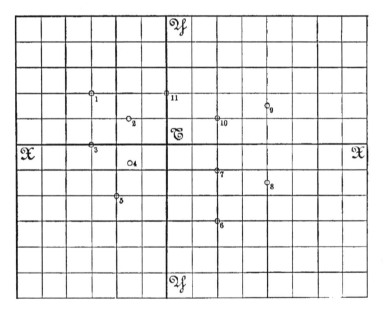

Fig. 4. — Determination of Telephonic Center. (An Odd Number of Stations.)

the abscissa 30.   Now, by reasoning in a similar manner to that used in the preceding example it is easy to see that if either axis, say, $Y\,T\,Y$, be shifted, say, one unit to the right and abscissæ to Nos. 1, 2, 3, 4, 5, and 11 will be *increased* one unit each, while the abscissæ to No. 6, 7, 8, 9, and 10 will be *decreased* one unit each.   Hence there will be a gross *increase* of six units and a gross *decrease* of five, or a net increase of one unit. The same argument applies equally to the other axis. So the shifting of either by any assignable quantity in either direction causes a positive increase in the sum of all the perpendicular distances of the subscribers from the axes.

It is now desirable to consider the application of this

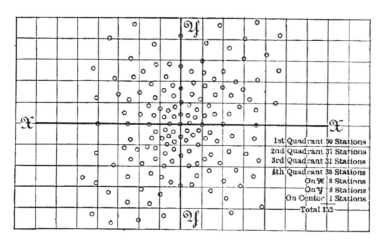

Fig. 5.—*Determination of Telephonic Center.* (*Average Case.*)

1st Quadrant	39 Stations
2nd Quadrant	37 Stations
3rd Quadrant	31 Stations
4th Quadrant	38 Stations
On X	8 Stations
On Y	8 Stations
On Center	1 Stations
Total	153

method to such distributions of subscribers as are commonly encountered, together with some special cases that are of interest. In Fig. 5 is given an example of the solution of the problem of finding the telephonic center under conditions that probably most frequently obtain; that is, a group of subscribers distributed over an approximately circular area in which the subscribers' density, or number of subscribers per unit of area, is greatest at about the center of the territory, and decreases with sensible regularity in all directions toward the cir-

cumference. In all the following examples the section
lines indicate the prevailing directions of the streets, and
consequently that of the axes, while the small circles
represent the subscribers. In this case there are 153 subscribers, and according to the preceding principles the
axis *y* must pass through one subscriber, and put 76 on
the left hand and 76 on the right hand. On trial, however, it is found that the nearest approximation that can
be made is to draw *Y* through nine stations, leaving 144

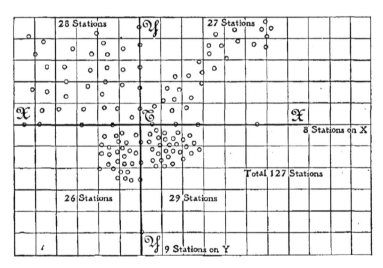

Fig. 6.—*Determination of Telephonic Center. (Irregular Distribution.)*

to be divided between the two sides, and the line is so
located as to place 72 on each side of it. The axis of *X*
is located in a similar manner. After the axes are drawn,
it seems that there are thirty stations in the upper left-
hand quadrant, five on the axis of *y* above the axis of *x*,
37 stations in upper right-hand quadrant, four stations on

x to the right of y, 31 stations in the lower right-hand quadrant, three stations on y below x, 38 stations in the lower left-hand quadrant, four stations on y to the left of x, and one station at the telephonic center.

Take the distribution illustrated in Fig. 6. Here the stations fall into four irregular lots — two small compact groups, and two in which the density is low, one nearly rectangular in outline and the other long and linear. There is a total of 127 stations. The nearest approximation that can be made is to draw y y so as to cut nine stations, putting 58 on the left side and 60 on the other, and x x to cut eight stations, placing 60 above this axis and 59 below, giving the telephonic center T at the intersection of the two lines. To test the correctness of this location, suppose y y to move one unit to the right, then the distance to the 58 stations to the left will be increased one unit. As the axis has moved one unit away from the nine stations that it previously cut, the mileage to each of them will be increased one unit, or there will be a total increase of 67 units. The motion of the axis will be toward the 60 stations on the right of it, hence there will be a decrease of 60 units on this side, or a total net increase of seven units. In the same manner, it is easy to see that if x x moves one unit, say upward, there will be an increase of seven units. Hence, neither line can move any assignable quantity without increasing the mileage.

When, therefore, such considerations as are thus illustrated are taken into account, the rule for locating the telephonic center is as follows: Draw a line through the group of stations parallel to one of the given directions, at such a location as will place as nearly as possible half the subscribers on one side and half on the other. Draw a second line perpendicular to the first at such a place as will again place as nearly as possible half the stations on one side thereof and half on the other. The telephonic center is at the intersection of the two lines.

Under certain distributions of stations, the location of the telephonic center will become partially indeterminate, or rather a number of sites may be found from any or all

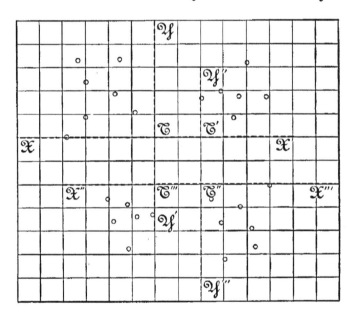

Fig. 7.—Determination of Telephonic Center. (Indeterminate Location.)

of which the wire mileage will be equal. Such a condition exists when there are no subscribers in the immediate vicinity of the center of the district. To illustrate, turn to Fig. 7. Here are 26 stations so distributed as to fall in 4 groups with no stations near the center of the territory.

Evidently the condition of placing one-half the stations on each side of each axis is fulfilled by drawing the axis of y anywhere between $y\,y'$ and $y''\,y'''$, and the axis of x anywhere between $x\,x$ and $x''\,x'''$, and any location of the telephonic center within the rectangle $T\,T''\,T'''\,T''''$ is equally good. To provide this, suppose the line $y\,y'$ to start from the position $y\,y'$ and move toward $y''\,y'''$. As there are 13 stations on each side of the line, every unit that the line moves to the right will add one unit to each of the lines to the stations on the left, and will subtract one unit from each of the lines to the stations on the right. As the number of stations is equal, equal amounts of mileage are added and subtracted, and the sum remains constant. Suppose, however, the moving axis in passage to the right *crosses* a station, *then* in the next unit that it moves, 14 mileage units will be *added*, and 12 units will be subtracted, hence there will be a net increase of two units. The same reasoning may be applied to the x axis. Hence the principle is established that there will be *no* change in mileage unless the moving axis *crosses* one or more stations. Thus if there are n stations in any territory, there will be $\frac{n}{2}$ stations on each side of each axis. Suppose a be the length of the average perpendicular distance from the axis to the n stations, then the sum of the total perpendicular distances to the axis will be

$$\frac{a\,(n)}{2} + \frac{a\,(n)}{2}.$$

Now, suppose the axis to move and cross m stations, then there will be $\frac{n}{2} + m$ stations on one side of it, and

$\dfrac{n}{2}-m$ on the other side. The mileage in this case will be

$$a\left(\frac{n}{2}+m\right) + a\left(\frac{n}{2}-m\right).$$

Subtracting this latter quantity from the first, it is seen

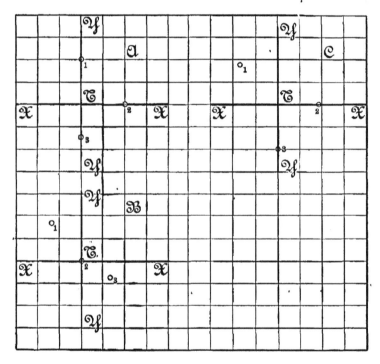

Fig. 8.—Determination of Telephonic Center. (*Special Cases.*)

that the resulting mileage is greater by 2 *am* than the former.

In Fig. 8 three examples are given of types of special cases that at first sight might seem to be exceptions to the rules as stated. At *a* there are three stations, or groups of stations. If $x\,x$ be drawn to place an equal number of stations on either side of it, it will pass through station 2, and if $y\,y$ be drawn perpendicularly to $x\,x$ and so placed as to give as nearly as possible an equal number of stations on each side of it, it will intersect 1 and 3. In this case the axes cut the three groups of stations. The total mileage is $T1 + T2\,T3$, and it is self-evident that if either axis be moved in either direction, this quantity will be increased. Turning to B, where the stations 1, 2, and 3 represent three isolated bunches of stations nearly in line with each other, if $x\,x$ and $y\,y$ be drawn to place equal numbers of stations on each side of them, and parallel with the section paper lines, they will intersect at 2, leaving 1 and 3 in diagonally opposite quadrants. Here T is at the telephonic center, and it is quite clear that any change in location thereof will augment mileage. In the distribution shown at C, if $x\,x$ be drawn to place an equal number of stations on each side of it, it will cut 2, and $y\,y$ under the same conditions will intersect 3, leaving 1 in the upper left-hand quadrant.

At first sight it might be supposed that the distribution of subscribers would be radically unlike in different towns or cities, but examination shows that the majority of cases, to say the least, will fall under six distinct and specific types of distribution. These six types may be denominated as follows:

1. *The Symmetrical System.* — In this type of distribution, the subscribers are spread over a territory approximately circular in contour, the greatest station density per unit of area being at the geometrical center of the district, and decreasing regularly radially in all directions.

This is the simplest, and probably the distribution most frequently met with, and is illustrated in Fig. 5. Examples of this type of distribution are so numerous that particular citation is superfluous, almost every small town or village affording an instance.

2. *The Isolated Group System.* — In this type there are two or more irregular groups of stations separated by ter-

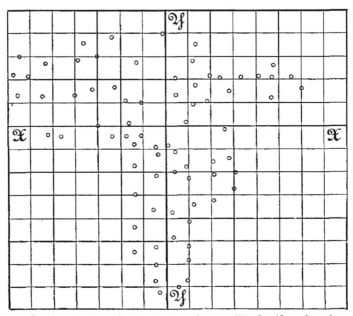

Fig. 9.—*Determination of Telephonic Center.* (*The Cruciform Group.*)

ritory telephonically sparsely populated. The station density is usually greatest at the center of each group, decreasing approximately radially toward the circumference of each. The usual form of this type of distribution is illustrated in Fig. 6, where there are four groups of stations, each entirely distinct from each other. The isolated group type of distribution is of very frequent occurrence. It is always met with where some natural obstacle, such as a river, divides a town into two or more parts, and is usual where a city is made by the absorption of several small towns. The upper part of New York City and the environs of Boston and Philadelphia are good examples. In Fig. 10, an extreme case of the isolated group type is given, such as would be presented by the establishment of an exchange to serve three small towns, or a group of

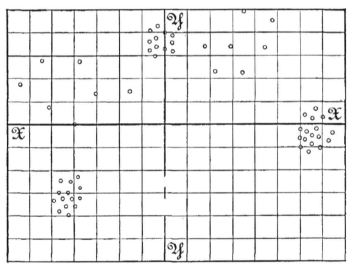

Fig. 10.—*Determination of Telephonic Center.* (*The Isolated Group.*)

large factories. The rule for selecting the telephonic center is perfectly applicable, and even though it indicates a spot that is apparently a telephenic desert, a little consideration shows that neither axis can be moved without increasing the total mileage. The next most common type is the *Cruciform System,* shown in Fig. 9. This distribution is always found where there are two principal streets at right angles to each other, a very common urban feature. In this case the axes are almost certain to lie either on or very close to the chief thoroughfares, and the telephonic center may, with reasonable probability, be located at the intersection of these streets without undertaking the task of map making or station plotting. The city of Chicago is an example of this type. In Fig. 11 the *Comet System* is exemplified. This type is of impor-

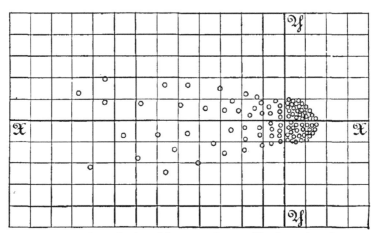

Fig. 11.—*Determination of Telephonic Center.* (*The Comet Group.*)

tance, as it is found in some of the largest cities, such as New York and Boston, though it is of less frequent occurrence than those previously mentioned. It is usually the result of geographical conditions, such as the building of a city on a peninsula, that forces all business interests into a small space at the apex of the peninsula, but allows residences to distribute themselves over the adjacent country. The territory is more or less fan shaped, or wedge-shaped with a group of stations of great density located at the small end of wedge; the density being maintained clear to the boundary lines, while toward the thick end, the station density very rapidly falls away, though the demand for telephones may extend many miles into the country.

The Crescent System is shown in Fig. 12, and in it, as its name indicates, the stations are arranged roughly in a

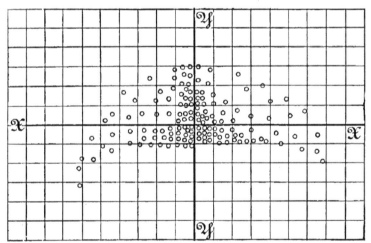

Fig. 12.— Determination of Telephonic Center. (The Crescent Group.)

lune-shaped figure, the density being usually greatest along a line through the center of the crescent. This is really a modification of the cruciform system. It is found in the lake and river cities, such as Cleveland, Milwaukee, Cincinnati, New Orleans, etc. Lastly, in Fig. 13, the *Linear System* is shown. Here the stations are scattered over a rectangular area much larger than it is wide, with the greatest density approximately at the center of figure. This type is of quite rare occurrence.

In the selection of the central office site it is impossible to give too much care and consideration to station distribution. If a telephone plant is in existence, its directory may be obtained and the subscribers plotted by dots on a large scale map. Such a representation gives a good idea of the arrangement of existing telephonic population, but the telephone engineer, more than any other mortal, lives in the future, and in order to make a successful

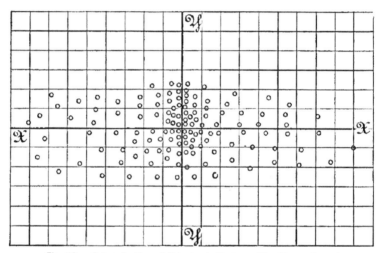

Fig. 13.—Determination of Telephonic Center. (The Linear Group.)

design, a reasonably accurate forecast both of the number and location of subscribers for a decade or more in advance must be made. To this end it is well to obtain the opinions of a large number of the oldest and most experienced residents, as to the probable growth both in amount and direction of the various kinds of business represented. For, in all towns, trades are clannish, and sooner or later districts spring up, and leather, law, dry goods, brokers, etc., segregate away from each other, each forming little communities by themselves, and the opinion of conservative representatives of each class as to its probable future furnishes the only basis on which a telephonic prediction can be made. It is particularly necessary to consider the effect of transit facilities, such as railways, freight stations, elevated and electric roads, etc., on the probable relation of the business and residence districts, and in this connection the steadying or anchoring effects of the modern office building is particularly noticeable. After the widest and most thorough canvass yielding the most complete obtainable information as to the expected future trend of the city under consideration is obtained, an estimate of probable subscribers may be made. The tendency is always to make such an estimate too low, because, for many reasons, telephonic business is growing faster than even its most sanguine promoters expect, so a liberal allowance should be made for future growth. Then, this estimated distribution should be plotted on a large scale map, from which the telephonic center may be located. In addition, it is most useful and instructive to construct a relief model of density and distribution. This is done by cutting out a board to scale, say, four inches equal one mile, representing the area under consideration. A number of strips of wood one-half an inch square, thus representing an area of 1-64 of a square mile, are prepared, and these strips cut of various lengths to represent the number of stations on each of the areas they represent. For example, suppose in the least dense portion 10 stations are estimated per block and in the densest portion 500 per block. Let the vertical scale be 100 subscribers per inch, then a piece of strip 1-10 of an inch thick will represent areas of minimum density, and a piece 5 inches will stand for the spots of greatest density. By thus cutting a piece of wood for each city block of such a length as to represent the expected number of subscribers on that block, and fastening it on the board in its proper place, a relief map of density and distribution is obtained that conveys the most vivid possible idea of the expected telephonic population.

CHAPTER VI.

OBSTACLES AND DIAGONAL STREETS.

I⊤ often happens that the topography of the territory under consideration prevents the complete utilization of the shortest rectangular routes to all stations, and the

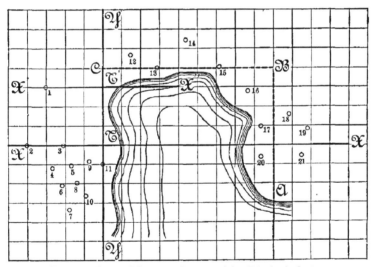

Fig. 14.—Effect of Obstacles in Determining Telephonic Center.

effect of such interference in causing circuitous routes must be taken into consideration in selecting the telephonic center. An example of such a case is given in Fig. 14. By the rule given under the hypothesis that all subscribers could be brought to the office by the shortest rectangular routes, the axes would be $x\ x$ and $y\ y$, with the telephonic center at T. But suppose that, owing to some unsurmountable obstacle, the stations from 12 to 21, inclusive, *must* be brought to T along the line $A\ B\ C$. Evidently, the effect of bringing stations 12 to 21, inclusive, to the telephonic center along $A\ B\ C$ is to deliver 10 lines at the point c on $y\ y$, and cause them to reach the office along $y\ y$. This is equivalent to moving them and concentrating them on the line $y\ y$ above $x\ x$. If, now, 10 stations are supposed to be at c, then the axis $x\ x$ must be shifted to $x'\ x''$ in order to fulfill the condition of equality in number of stations on each side of it. The true telephonic center is then at T' instead of T.

In many cities, particularly the older ones, diagonal streets are of occasional occurrence that may be employed for wire routes, and produce a saving in mileage over that required by running strictly rectangularly. While the advantageous use of such streets is infrequent, it is necessary to consider their effect on the location of the telephonic center in order that a complete solution may be found.

When a diagonal street can be employed, that portion of each subscriber's line which follows the diagonal street becomes approximately an air line, and is no longer referred to rectangular co-ordinates. Then each circuit will either be measured all rectangularly, all straight, or partly rectangularly and partly straight. This mixing of the methods of measurements so complicates the problem that no direct mathematical solution has been discovered, and it is only possible to ascertain the telephonic center by the method of trial and approximation, best accomplished by reference to a map on which the station locations are plotted. A concrete example will elucidate this method. In Figs. 15 and 16, let the line *A B* represent any diagonal street intersecting the co-ordinate axes at any angle whatever extending any length. Let *T* be the telephonic center referred exclusively to the rectangular axes. Simple inspection shows that the stations which lie closely to *A B*

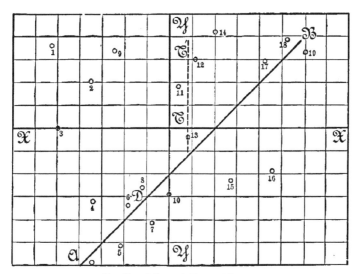

Fig. 15.—*Effect of Diagonal Streets.*

can reach *T* with the expenditure of less mileage by following the line *A B* till it intersects either of the axes, and then proceeding along the axes, than by taking a purely rectangular course. But it is a question whether a new location for *T* at some other point would not still further economize mileage. Examining Fig. 15 it is seen that there are 19 stations, of which eight, Nos. 4, 5, 6, 7, 8, 17, 18, and 19, can shorten their wire mileage by taking the street *A B*, while for the remaining 11 the rectangular route would be the shortest, so there will be eight stations that can employ *A B* to advantage and 11 that cannot. For these 8 stations that use *A B*, the location of the office on this line will give shorter wire routes than any point outside the line. If the office is not placed on *A B*, then these 8 stations must have their lines lengthened to reach it at some other point. But there are 11 stations that

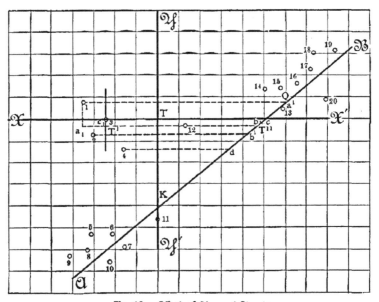

Fig. 16.—Effect of Diagonal Streets.

could not use *A B*, and for which the best office location would be somewhere outside *A B*. Hence, if 8 stations have their wire mileage increased by moving off *A B*, and 11 have their mileage decreased, it will not be economy to locate on *A B*. Turning to Fig. 16, it is seen that there are 20 stations, of which 14 could use $A\ B$ as a portion of their wire route. Evidently in Fig. 16, the conditions of Fig. 15 are reversed, that is, a greater number of stations could use the diagonal street $A\ B$ than would reach the point T rectangularly. Therefore, the effect of diagonal streets on the office location depends only on the question as to whether a greater number of stations can use the diagonals to advantage than can reach the office rectangularly.

To ascertain this condition locate the telephonic center for rectangular co-ordinates and draw the axes, then

1. If diagonal streets are found cutting only one of the quadrants, the use of such streets will not affect the office location. For (refer to Fig. 17) let $A\ B$ be the direction of any diagonal street in one quadrant making an angle a with the line $A\ C$, which may be either the axis of x, or y, or a line parallel to either, and assume the office to be at A, or anywhere to the left and above the point

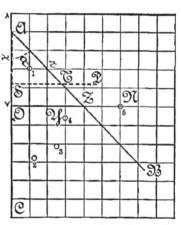

Fig. 17.—*Geometrical Construction for Diagonal Streets.*

A. If any station, or cluster of stations, at any point N be brought to the office by rectangular co-ordinates, the location of the wire route will be from N to O, thence from O to A, and thence to the office, or by some equivalent rectangular route. If the diagonal street $A\ B$ be used as a portion of the route, the line will run from N to Z, from

Z to A, and thence to the office. From A to the office the length of the route is not affected by the use of the street $A\,B$, hence it is only necessary to deal with such a part of the route as is to the right of and below A. The distance from N to A measured rectangularly is

$$AO + ON.$$

Let

$$X = AO \quad \text{and} \quad Y = ON.$$

Then the distance

$$r = AZ$$

(the portion of the diagonal street occupied) is given by the expression

$$\frac{X}{\cos a}$$

and the distance zn by $y - \dfrac{X}{\sin a}$,

and the total distance from A to N using $A\,B$ is

$$\frac{X}{\cos a} + y - \frac{X}{\cos a}.$$

If the office is not at A, but at some other point distant rectangularly D miles from A, the total mileage for the group N from the office will be

$$D + \frac{X}{\cos a} + y - \frac{X}{\sin a}.$$

Turning back to Eq. 3, it is evident that any term in this equation may be replaced by the preceding expression for station mileage *via* diagonal streets, and if the same mathematical reasoning and processes be applied to the equation so obtained, the same condition is found to hold true for U and V, hence the location of the office is not changed by the employment of diagonal streets in this case.

To determine the distance from the office of such stations as can use diagonal streets, recourse may be made to the map on which the stations are plotted, and scale measurements made from the office to A, thence from A to Z, and thence from Z to N. If desired, the following geometrical construction may be employed to locate such stations as can make use of diagonal streets in such a hypothetical situation as will make their rectangular coordinates in their new position equal in length to the mixed rectangular-diagonal distance of their true position. From the point A lay off $A\,S + S\,T$ equal to $A\,Z$. From the point T draw $T\,P$ parallel to the axis of y and equal to $Z\,N$, then the sum of the distances $P\,S + T\,S + S\,A$ is equal to $A\,Z + Z\,N$. Therefore, so far as length of lines is concerned, it is allowable to assume the group N as located at P, and to measure the distance to the office rectangularly along $T\,S$ and $S\,A$, for this gives identically the same length as $N\,Z$ and $Z\,A$.

2. If diagonal streets be found passing through two quadrants, it is possible, though not probable, that the utilization of these streets will change the location of the telephonic center. The method of determining this is given under case third below.

3. If diagonal streets be found passing three quadrants or more, it is probable, though not certain, that the employment of these streets *will* change the location of the telephonic center. To determine whether the office is

moved to a new position, turn back to Fig. 16. Using
rectangular co-ordinates, the telephonic center is at *T*.

By inspection, it is seen that stations 5, 6, 7, 8, 9, 10,
13, 14, 15, 16, 17, 18, 19, and 20 can avail themselves of
the line *A B*, while stations 1, 2, 3, 4, 11, and 12 would
use rectangular co-ordinates to reach *T*. Consider first
the stations that can use the line *A B*. The location of
the office for these will be somewhere on the line *A B*, at
such a point as to numerically bisect the number of stations that follow this route. As there are 14 stations that
can use *A B*, the office will be located between the seventh
and eighth station, or at the point *Q*. Inspection shows
that if *Q* be moved to cross station 13, then for every
unit it moves toward *A*, the lines to eight stations will be
increased and to six stations decreased. By the same
reasoning, if *Q* moves across station 14, the same condition will prevail; therefore, so far as these stations are
concerned, the office must be placed at *Q* between stations
13 and 14. Turn now to stations 1, 2, 3, 4, 11, and 12.
It is possible to locate the proper telephonic center for
these stations considered by themselves, which is found to
be at *T"*. It is now necessary to combine the points *T"*
and *Q* to obtain the true minimum mileage. As there are
14 stations that can use *A B* advantageously, and only six
that cannot, it is probable that the best office location will
be somewhere on *A B*. Suppose each of the stations 1,
2, 3, 4, and 12 be brought to *A B* by straight lines parallel
to *X*, as shown by the dotted lines, to the points *a b' b c
a'*, while station 11 is brought along *Y T Y* to *k*. It has
been shown that the proper office site for the stations 5,
·6, 7, 8, 9, 10. 13, 14, 15, 16, 17, 18, 19, and 20, is at *Q*,

but by the introduction of stations 1, 2, 3, 4, 11, and 12, five additional lines will run toward Q from A, and one from B, so, now, there will be 12 lines to Q from A, and 8 from B. Hence Q must move toward A till it crosses two stations, thus numerically bisecting the total number. This will be at the point T''', between the lines from stations 3 and 12. Whenever, therefore, diagonal streets arise, the three preceding criteria may be applied to determine whether the use of such streets will cause the office to leave the telephonic center of rectangular coordinates. As the application of these criteria is a matter of some considerable labor, in a large group of stations, it is usually possible to decide, from a careful inspection of a properly prepared map, whether such departure will be of sensible magnitude, and of sufficient importance to justify spending the time needed for exact location by the preceding method of trial and approximation.

CHAPTER VII.

THE EFFECT OF MOVING THE TELEPHONIC CENTER.

IT is often impossible, and sometimes inexpedient, to place the office exactly at the theoretical telephonic center, so it is necessary to consider the effect on the expense of the wire plant of a departure from the true location. The removal of the office in any direction is equivalent to shifting either, or both, of the co-ordinate axes. Let S be the total number of stations in the territory, and L the average rectangular distance to all stations, so that $S L$ is the total wire mileage to reach all stations from the telephonic center. Now, suppose the axis of y to move parallel to itself to the right hand away from the telephonic center T to a new position T' and let L_x denote the distance that the axis moves. Originally, there were $\frac{s}{2}$ stations on each side of this axis. Owing to the motion of the axis, each of $\frac{s}{2}$ stations on the left of the axis will have its line increased an amount equal to L_x units in a direction parallel to the axis of X, and there will be an increase in wire mileage of $\frac{s}{2} L_x$ units. As the axis of Y moves to its new position T'', it will pass over, or cross, a certain number of stations. Or, in other words, there will be a certain number of stations included between its first and second locations. Let S'_m stand for the number of these stations. Then, on one side of the axis of Y in its new location there will be $\frac{s}{2} + S_m$ stations, and on the other $\frac{s}{2} - S_m$ stations. Owing to the motion of the axis this last number of stations will each lose L_x units of mileage, or there will be a decrease of $L_x\left(\frac{s}{2} - S_m\right)$ miles.

Now, to ascertain the *effect* of the moving axis on the mileage of the stations that it crosses during its motion, refer to Fig. 18. Let T be the true telephonic center, $Y\,T\,Y$ the axis of Y through that center, and $Y'\,T''\,Y'$ the axis of y in a new position through T' distant L_x units from T. Let S be any station distant x units from $Y\,T$ Y that is crossed by the axis of Y moving to its new position. Then S, when referred to $Y'\,T''\,Y'$, will have a new

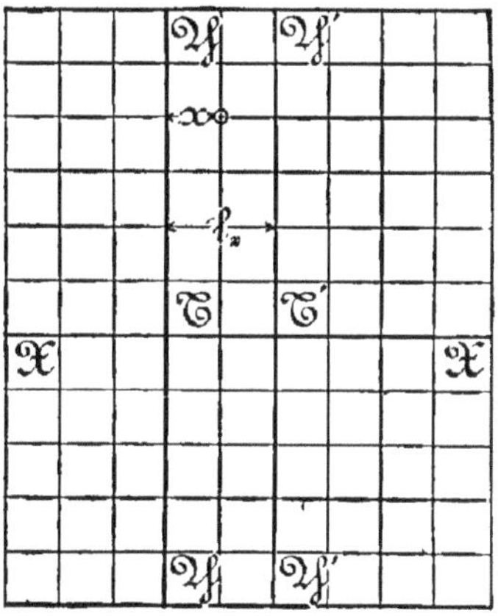

Fig. 18. — *Effect of Moving Telephonic Center.*

abscissa equal to $L_x - x$, and the difference between the new and old abscissæ, for S will be $(L_x - X) - x = L_m - 2x$. If there are m groups of stations between $Y\,T\,Y$ and $Y'\,T''\,Y'$ having abscissæ varying from x to x_m, these stations will have their mileage increased by an amount equal to

$$S_1 (L_x - 2\,x_1) + S_2 (L_x - 2\,x_2) + S_3 (L_x - 2\,x_3) + ''''' S_m L_x - 2\,x_m.$$

Then, with reference to the new axis of Y the total abscissæ of all stations will be increased by the following quantity:

$$\frac{S}{2}L_x - \left(\frac{S}{2} - S_m\right)1\,x + S_1(L_x - 2\,x_1)$$
$$+ S_2(L_x - 2\,x_2) +''''S_m(L_x - 2\,x_m).$$

Reducing,

$$S_m L_x + S_1(L_x - 2\,x_1) + S_2(L_x - 2\,x_2 +'''' S_m(L_x - 2\,x_m).$$

In this expression, S_1 to S_m are the stations crossed by the moving axis. The same reasoning may be applied to

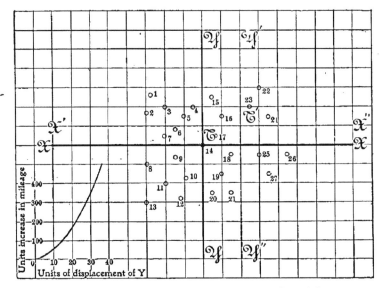

Fig. 19.—*Effect of Moving Telephonic Center.* (*Example.*)

the axis of X; hence, to ascertain the effect of a change of location in either axis, it is only necessary to consider the amount that the moving axis will *change the perpendicular distances of the stations that it crosses.* If both axes move, the full solution is obtained by adding the partial solution for each axis. For any given case, therefore, it is easy to imagine either or both of the axes shifted such varying amounts as will cover all central office sites worthy of consideration, then by substituting numerical values in the preceding expression a curve may be easily plotted, showing the relation between distances from the true telephonic center and increase in line mileage.

In Fig. 19 a concrete example is given. There are 27 stations. The axes are shown by the lines $X\,X$ and $Y\,Y$ cutting station 14, leaving 13 stations on each side of each axis, giving the true telephonic center at T. Now, assume Y to be shifted 20 units to the right, and X 10 units upward, as shown in dotted lines, to what extent is mileage increased. In table No. III. (page 84) the perpendicular distances of all stations from both the true telephonic center and T and the displaced one T' are given. From this table the X ordinates are shown to be increased 32 units and the Y ordinates 166 units, a total of 198 units. Applying the formula just developed to determine the increase in mileage due to moving the X axis gives the following:

$$S_m = 3$$
$$L_x = 10 \qquad S_m L_x = 3 \times 10 = 30$$

$$S_6\,(L_x - 2\,x_6) = 1\,(10 - 2 \times 9) = -8$$
$$S_7\,(L_x - 2\,x_7) = 1\,(10 - 2 \times 9) = 0$$
$$S_{14}(L_x - 2\,x_{14}) = 1\,(10 - 2 \times 0) = 10$$
$$S_{17}(L_x - 2\,x_{17}) = 1\,(10 - 2 \times 5) = 0$$
$$\overline{+\,2}$$

$$S_m 1_x \qquad\qquad\qquad\qquad 30$$

Total net increase for x . . . . $\overline{32}$

TABLE No. III.

Effect of Shifting Telephonic Center in Fig. 19.

STATION NUMBER.	DISTANCE FROM			
	$x x$	$y y$	$x' x''$	$y' y''$
1	26	28	16	48
2	17	30	7	50
3	20	20	10	40
4	20	5	10	25
5	15	10	5	30
6	9	15	1	35
7	5	20	5	40
8	10	30	20	50
9	6	15	16	35
10	17	9	27	29
11	20	20	30	40
12	28	12	38	32
13	30	30	40	50
14	0	0	10	20
15	25	5	15	15
16	15	10	5	10
17	5	7	5	13
18	5	15	15	5
19	15	10	25	10
20	25	5	35	15
21	25	15	35	5
22	30	30	20	10
23	20	25	10	5
24	15	35	5	15
25	5	30	15	10
26	5	45	15	25
27	15	35	25	15
Totals	428	511	460	677
Deduct	. . .	. . .	428	511
Differences	. . .	. . .	32	166
Add	. . .	. . .	. . .	32
Total increase	. . .	. . .	. . .	198

The increase due to the shifting of the axis of y is calculated as follows:

$$S_m = 7 \qquad L_y = 20 \qquad S_m L_y = 7 \times 20 = 140$$

$$S_{14} (L_y - 2\, y_{14}) = 1\,(20 - 2 \times \ \ 0) = 20$$
$$S_{15} (L_{y15} - 2\, y_{15}) = 1\,(20 - 2 \times \ \ 5) = 10$$
$$S_{16} (L_y - 2\, y_{16}) = 1\,(20 - 2 \times 10) = \ \ 0$$
$$S_{17} (L_y - 2\, y_{17}) = 1\,(20 - 2 \times \ \ 7) = \ \ 6$$
$$S_{18} (L_y - 2\, y_{18}) = 1 \cdot (20 - 2 \times 15) = \quad -10$$
$$S_{19} (L_y - 2\, y_{19}) = 1\,(20 - 2 \times 10) = \ \ 0$$
$$S_{20} (L_y - 2\, y_{20}) = 1\,(20 - 2 \times \ \ 5) = 10$$
$$S_{21} (L_y - 2\, y_{21}) = 1\,(20 - 2 \times 15) = \quad -10$$

$$46 - 20 = 26$$
$$140$$

Total 166

Increase for shifting of axis of x 32

Total increase (check with Table III.) . . 198

In the lower left-hand corner of Fig. 19 a curve is given, showing the relation of increase in mileage and displacement of the axis of y in the example of Fig. 19. The vertical scale is change in wire mileage, while the horizontal scale is units of displacement. A similar curve could be plotted for the motion of the x axis, while a combination of the two curves would give the resultant of a change in both axes. To completely cover the problem it is necessary to plot a curve for each axis in order to show the effect of motion in each direction. While from the formula and specimen curve of Fig. 19 it is easy to see that, in general, the curve of mileage increase will be parabolic or hyperbolic, differing station density will so change the constants that no general curve can be given, and it is necessary to plot a specific one for each case.

PRACTICAL LIMITATIONS.

While the preceding discussion indicates the method of selecting the central office site, in order to make the wire plant expense a minimum, there are certain practical limitations due to urban topography that must not be completely ignored. The underlying principles can be best illustrated by a concrete example, and as there is no exact mathematical treatment each case must be worked out by itself. In Fig. 20 let the plane of the paper represent any town laid out in approximately rectangular blocks, with 27 stations indicated by the small circles. By the method described the axes will be located at XX' and YY', with the telephonic center at T. On the left-hand the streets horizontally are designated by letters from A to E inclusive, while vertically along the top they are numbered from 1 to 6 inclusive. Theoretically, the wire plant is supposed to extend from the switch-board to the sub-station instrument, practically it would run from the office manhole in the street, near the wall of the office building, to the wall of the subscriber's premises; the inside wiring of both sub-stations and office being excluded from the definition of wire plant. Theoretically, each line is assumed to follow the shortest rectangular route from the office to the sub-station, but practically this is not always possible. Thus, station No. 21 should theoretically proceed west to the line YY, and thus north to the center at T. Practically, station 21 must proceed east to Sixth Street,

thence north to C Street, and hence west to the office.
This course is necessitated, as it is ordinarily impractical
to obtain access from the street to the premises of a station,
excepting through the owners' frontage.   Similarly stations
4 and 6 should run directly east to the axis YY, and
thence south to the office location T.   But this is impractical, as they are located in the center of a block, and

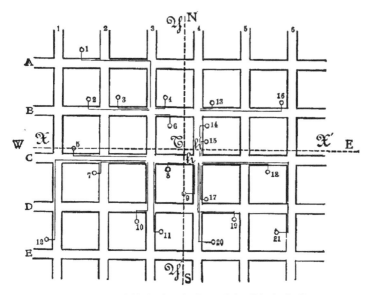

Fig. 20.—Practical Limitations in Determining Telephonic Center.

cable runs would not be constructed through private
premises to reach the office ; thus all lines must follow
the shortest *practical* routes to the nearest street, and then
be carried through conduits or on pole-lines to the office.
From inspection of Fig. 20 it is seen that the nearest
street points to the telephonic center T, where an office
manhole could be constructed, are the points H and H',

and that either of these locations is equally good, so far
as length of cable from the manhole to the office is concerned, and it becomes only a question of which of these
two points will give the shortest wire routes to the stations.
Further inspection shows that for subscribers 4 and 6,
there is very little choice between running east on B Street
and south on Fourth Street to H' or west on B Street to
Third Street, thence south to C Street and east on C street
to H; for subscribers 14 and 15 the route on Fourth
Street to H' is a little shorter than the mileage required
to reach H, and the same is true for station 8. If H is
used for the office manhole, the line to station 8 is shortened, while from stations 14 and 15 the lines are lengthened. Contrariwise, if H' is used, the lines of 14 and 15
are shortened and to 8 lengthened.

Considering the whole of the territory, it is seen that if
all the stations on the west of the axis of Y, except 8, are
brought to the intersection of Third and C streets, there
will be five lines running south on Third Street, three
lines east on C Street, and two north on Third Street, a
total of 10 lines converging to this point. On the east
side of the axis of Y, to the intersection of Fourth and C
streets, there will be four lines proceeding south on Fourth
Street, two west on C Street, and four north on Fourth
Street, a total of 10 lines. If station 8 be included there
will be 11 lines from the west and 10 from the east.
Similar reasoning may be applied to the stations on each
side of the axis of X. Now, it is evident that the office
manhole may be moved north or south on Third or Fourth
streets, or east or west on B or C streets, without materi-
ally affecting the total mileage, so long as it is not carried

beyond the rectangle formed by these streets. It is also a matter of indifference on which side of these streets the office is located. Hence the practical rule that the office may be located anywhere within a rectangle formed by the streets bounding the block in which the theoretical center is found.

CHAPTER VIII.

THE RELATION BETWEEN SUBSCRIBERS' DISTRIBU-
TION AND WIRE MILEAGE.

ATTENTION has already been directed to the effect of
the distribution of the subscribers on the location of the
central office, and it is now necessary to consider, in addition to the shape of the group of sub-stations tributary to
the office in question, the density of the group, or the
number of sub-stations, per unit of area ; the way in which
changes in density affect the cost of installation, and the
complete relation between the shape of the group, its density and the necessary wire mileage to reach all stations.
The simplest case is that of a group of regular geometrical
section, such as a square, or circle, over which the sub-
stations are distributed in uniform and regular order.
Unfortunately, such a group never occurs in practice, but
even the most heterogeneous arrangement may, by differentiation, be split up into sensibly uniform parts, and the
results of reasoning applied to each differential will, when
integrated, supply a solution for the whole area. Refer-
ring to Fig. 21, suppose the territory under consideration
to be represented by the square $a\,b\,c\,d$, over which the
stations are uniformly distributed. Let l be the length of
one side of the square, then the area will be l^2, and if there

are S stations per unit of area, the total stations will be Sl^2. The telephonic center will evidently be at the geometrical center of the figure, and as all stations are brought to the office by rectangular routes the best wire plant system will consist in building a main lead along a diameter of the area passing through the office, say, at $y\,y$, and bringing to this main lead all the sub-stations lines, as

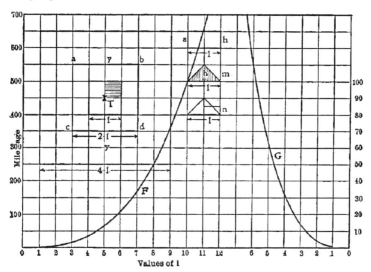

Fig. 21. — Grouping of Sub-stations.

shown by light black lines in the upper right-hand quadrant, and then carrying them to the office along the main lead. If l be the side of the square, it is evident by inspection that the average distance of all the stations from the main lead is $\dfrac{l}{4}$, and the average length of each line along the main lead is also $\dfrac{l}{4}$, hence the average distance of all stations from the office is $\dfrac{l}{2}$, and the total distance or circuit mileage of all stations is $Sl^2 \times \dfrac{l}{2} = \dfrac{Sl^3}{2}$. For me-tallic circuits the wire mileage will be double the circuit mileage or Sl^3, thus giving a very simple expression for the total wire mileage or *lineage*, as this quantity may for brevity be called.

If a plane be imagined to be passed along $y\,y$, the dis-tribution of sub-stations will be represented by the pro-jection of this plane on the plane of the paper as shown at *II*, in which S represents to scale the sub-stations per unit of area and l the length of the section of the group. Suppose now the group to increase in size until l becomes $2l$, $4l$, etc., the area will be 4 times, 16 times, etc., as large as the original area, then the successive total circuit mileages will be found (by substituting $2l$, $4l$, etc., for l) to be $4\,Sl^3$, $32\,Sl^3$, etc., but in each of these expressions l is the length of the side of the original plot. As S or the number of stations per unit of area is a factor of the first power in the expression, any change in the station density will simply make a pro rata change in the wire mileage. The curve F given in Fig. 21 gives the relation between the total circuit mileage and the size of the terri-tory covered. The vertical and horizontal scales must be read in the same units; thus if l is taken in miles, then the left-hand scale giving the circuit mileage must be read in miles, and the station density is to be taken in the same unit squared. For example, suppose a certain territory to be two miles square, and to contain one subscriber per square mile, then from the curve four circuit miles or eight wire miles will be needed to reach these stations. Assume another group four 1000 ft. on a side, with one station per 1000 ft. square, then the circuit mileage will be 32 × 1000 ft., or about six miles, and the wire mileage about 12 miles. If S or the number of stations per unit of area be different from one, it is only necessary to multiply the mileage thus found by the value of S. Thus in the preceding example, suppose S to have a value of 5 per 1000 ft. square, then the needed circuit mileage will be 32 × 1000 × 5, or about 30 miles. At the right hand of the sheet the curve G is a portion of the lower part of the curve F magnified five times for convenience in reading small value of circuit mileage.

Suppose now the shape of the group is that of a rectangle the sides of which are a units and b units, respectively, then the area will be ab and the total number of stations Sab. If the main lead is built parallel to the side a, then the average circuit distance to the main lead will be $\frac{b}{4}$ and the average length of circuit along the lead $\frac{a}{4}$. The total circuit mileage will be $Sab\left(\frac{a+b}{4}\right)$ and the wire mileage $Sab\left(\frac{a \times b}{2}\right)$. An expression of the same value is easily found if the main lead is built parallel to the side b, showing that either diameter may be used indifferently to give the same total wire mileage. An extension of the same reasoning may be applied to a group of any shape, though those which are, or those which approximate to, the well-known geometrical forms, are easiest dealt with. .

So far the reasoning has proceeded on the assumption that the density of the distribution of sub-stations was uniform over all the area under consideration. But this condition rarely if ever obtains; on the contrary, station density per unit of area is usually greatest near the telephonic center, and decreases with greater or less regularity towards the perimeter of the area. To consider the effect of a varying density return to Fig. 21, and again consider the square $a\,b\,c\,d$ to be the territory under consideration, but suppose the density to be a maximum of, say, h at the point T, and to decrease regularly to the bounding lines in all directions. Imagine now, for the sake of obtaining a clear mental picture, that the stations are piled on top of each other all over $a\,b\,c\,d$ in proportion to the assumed density. This will give a solid of the shape of a regular four-sided pyramid, of which the base is the area $a\,b\,c\,d$ and the altitude is h, or the maximum station density. If a plane be passed along y, the section of the pyramid will be a triangle with a base l and an altitude h. The vertical lines or ordinates to the base l shown at m (Fig. 21), represent the varying station density, while the horizontal lines or abscissæ to the altitude of this triangle are proportional to the circuit mileage required to reach the several clumps of stations that make up the entire group, and the product of all the ordinates and abscissæ will be proportional to the circuit mileage required to bring the stations to the office, or, in other words, to the area of the triangle. But this area is equal to that of a rectangle of the same base, and half the altitude; hence it is proper to replace a non-uniform distribution by a uniform one of the same base and a proportional altitude, and calculate lineage by the preceding method. So all forms or densities of sub-station groups may be reduced to equivalent simple areas and densities, and considered accordingly.

The slightest inspection of the curves of Fig. 21 shows that the necessary lineage to reach sub-stations increases very rapidly as the size of the territory tributary to any office is augmented, the formula for mileage showing the ratio to be proportional to the cube of the side of the territory, and directly proportional to the station density. In formulating, therefore, the design of any telephone plant, the present and future station density, and present and future area to be served, must be carefully considered separately from each other. If further future growth is to take place by increments in density, the wire plant expense will increase pro rata with the density, while if growth in territory be anticipated, the wire plant cost will grow as the cube of the side of the original area. Ordinarily it is probable that there will be both an increase in station density and territorial expansion, though as telephone companies usually work within well defined boundaries, specified by franchise grants, such as town or city limits, the ultimate available territory is usually well defined and grows very slowly, only as fast as corporate limits are extended, yet it is rarely that a telephone company plans at its start to cover all territory included in its franchise.

THE AVERAGE LENGTH OF LINE.

If in any territory the total circuit mileage needed to reach all sub-stations be divided by the total number of stations, a quantity is obtained that represents the *average circuit mileage* of each station from the office, or the average distance of all sub-stations. For a square group with uniform distribution the average mileage is by the previ-

ous discussion $\frac{l}{2}$, for a rectangle $\frac{a+b}{4}$. For all non-uniform station densities, the average mileage will be less
than the preceding expressions, except in the extraordinary
case of a density greater at the perimeter than at the
center. Though average mileage is an implicit function
of density and size of territory, its dimensions do not
convey the slightest idea of either of these quantities, for
a small exchange with low density may yield an average
mileage either greater or less than a much larger one with
greater density. While a theoretical investigation of total
and average mileage is an interesting subject, the practical
telephonist is usually more concerned in what *is* than as
to *why* it is, and the actual determination of average mileage is an easy though somewhat laborious task. For an
untelephoned town the most accurate possible estimate
obtainable is made by plotting on a good map the location
of probable sub-stations, and then making measurements,
following streets, from the office site to all stations.
As, however, a large number of cities possess telephone
plants, it is easy, by plotting from the telephone directory
the location of all sub-stations, to calculate the average
existing mileage with great accuracy. By thus tabulating
from a large number of cities a valuable collection of data
may be gained from which it is possible to prophesy with
more or less accuracy the probable average mileage in a
similar untelephoned town. There are, however, at present two perturbing influences, which should cause deductions from such source of information to be received and
used with the greatest caution. The first of these disturbing causes lies in the rapid increase in the number of sub-

scribers to all existing telephone plants. This increase
arises partly from the natural growth of the business,
partly from unprecedented prosperity and business activity
in the whole country, but at present chiefly from a keen
competition that stimulates all telephone companies to
secure as many subscribers as possible, by all means in
their power, thus incidentally lowering telephone rentals
and bringing service within the reach of many who other-
wise would not subscribe. From these causes it is safe to
estimate that within ten years past the telephonic population of American cities has more than quintupled. The
general effect of this increase in sub-stations is to decrease
the average length of line, though this decrease is by
no means proportional to the rate of increase in numbers.
For with a diminishing telephone tariff, there is a tendency
to build up the station density about the perimeter of the
territory inhabited by the less wealthy classes, much more
rapidly than in the center, where the converse holds true.
Again, in many cities two competing telephone companies
are established. There are some who subscribe to one
company, some to the other, and some to both. While as
a matter of fact it is easy to determine the actual average
length of line by dividing the total wire mileage by the
total number of sub-station instruments, such a deduction
is misleading, as telephonically two instruments at the
premises of one subscriber are superfluous. With these
reservations the curves shown in Fig. 22 may be considered as approximate indications of the probable average
length of line to be expected. It is convenient to divide
these curves into three groups : Those that apply to large
cities having a population over 500,000, those for medium

cities with, say, 200,000 to 300,000 inhabitants, and those
for small cities, having 100,000 inhabitants or less. The
curves *a*, *b*, and *c* apply to the first group. On the lower
horizontal scale the number of sub-stations included in the
office will be found, while on the right-hand vertical scale
the corresponding circuit mileage in tenths of a mile is
given. The curve *a* refers to the central portion of a

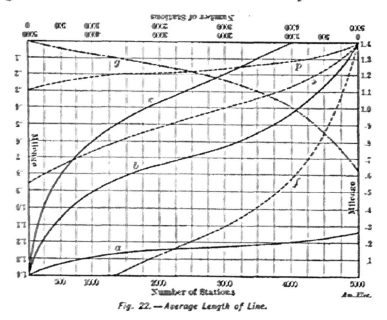

Fig. 22.—Average Length of Line.

large city, curve *b* to a part midway between the center
and the outskirts, and curve *c* to the outer portion. The
curves are used as follows: Suppose it be desired to find
the probable length of line in the middle portion of a large
city for an office of 2500 subscribers. On the lower
horizontal scale find 2500, then follow a vertical until the
curve *b* is intersected, then a horizontal to the right-hand

scale reading .705 miles. The curves *d*, *e*, and *f* are similarly the data for the central, middle, and outer portion of a medium-sized city, while the curve *g* applies to the center of a small city or a large town. To avoid confusion these curves are plotted to scales on the top and left hand of the sheet, which should be turned over to read the curves. It should not be forgotten, however, that these values apply only to *wire plant used*, and as a certain surplus must always be counted on, the curve values must be multiplied by from 50 per cent to 100 per cent if they are used to indicate the probable plant necessary to install.

CHAPTER IX.

DECREASE IN LINEAGE BY SUB-DIVISION OF TERRITORY.

An inspection of the curves of Fig. 21 shows that if any given territory be sub-divided into several parts, and an office placed in each one, the sum of the wire mileage needed to serve all the stations in the sum of the parts, is much less than that required when the territory is treated as a whole tributary to one office. Thus, for example, a square territory divided into four parts will need one-half the mileage of the same area taken as a whole, and the question at once arises, why not sub-divide all territories into a large number of small offices? This at once introduces the great debatable ground in telephony, of one office versus many offices, over which the entire fraternity have from the beginning been divided into two opposing factions, and about which has been fought a long, wordy war, with here and there spits of bitterness, far more indicative of personal prejudice than of such a calm spirit of impartial evaluation of evidence as is demanded for the proper investigation of scientific or economic problems. Like many other controversies, however, this question resembles the old argument about the Druid's Shield, for there are two entirely distinct and totally different sides to it, and the controversants rarely take the pains to consider that the color of the other side may be entirely different from the one at which they are gazing.

There is on the one hand the economic or engineering aspect that considers only how to build and operate the best plant for the least money, while on the other is the standpoint of character of service rendered to the subscriber; a question belonging strictly to the general manager, being entirely foreign to the province of the engineer; unless, as is sometimes the case, a single individual attempts, " Poo Bah " fashion, to combine both offices. It is beyond cavil that within reasonable limits the sub-division into several offices of a territory either containing a large number of stations or spread over a territory of considerable area, will decidedly cheapen both the cost of original construction and that of annual operation over the same territory designed on the one office plan. But it is equally *sure* that with this multiplicity of offices the *service* to the subscriber will be somewhat slower, and the percentage of errors increased. It is the province of the engineer to estimate as accurately as possible the probable cost of equipment and operation on both methods, and express an opinion as to the relative excellence of the service rendered by each. But here the engineering questions end, for it is purely a general manager's question to decide whether the difference in service is worth the cost thereof, and whether the finances of a particular company will permit the expenditure of the required difference in investment and annual expense. A precisely similar problem is presented in the determination of the kind of switchboard system to be employed.

A well constructed magneto exchange is cheaper to install
and maintain than a common battery system, but the service is not so pleasing nor effective. To telephonic
engineering belongs the problem of estimating the relative
cost of each, the determination of the kind of service each
will give, and the successful installation of whichever
system may be selected, in the cheapest possible manner
consistent with good work. But the selection of the one
to be used in a given instance is a general manager's
problem, with which the engineer has nothing to do. In
the succeeding paragraphs, therefore, it is proposed to
detail the effect on cost of installation and operation of
the sub-division of a given territory into several offices,
and the effect of this sub-division on the service rendered,
as compared with a single office, and leave the reader to
form his own estimate of the relative value of these two
incommensurable quantities — quality of service and cost
of plant — for there is no common unit by which both can
be measured, and this is the real root of all differences
of opinion on this question. But first, however, it is
necessary to exhibit the method of sub-dividing any
territory into subordinate parts in such a manner as to
demand the minimum wire mileage to reach all sub-
stations in each division.

THE SUB-DIVISION OF A GIVEN TERRITORY IN ORDER TO MAKE SUB-STATION MILEAGE A MINIMUM.

In order that any territory may be divided into several
subordinate parts such that the sum of the mileage in all
the parts to their respective sub-stations shall be a minimum, two conditions must be fulfilled. These are:

Condition I. — The office in each division must be at the telephonic center of that particular division.

Condition II. — Each division line between each of the sub-divisions must bisect and be perpendicular to a line joining the telephonic centers of the two groups that it separates.

CONDITION I.

It has already been shown that for any arrangement of sub-stations a minimum mileage is secured by locating the office at the telephonic center. If any territory be sub-divided into a number of smaller parts, and the office of each part be located *away* from the telephonic center of that part, the mileage for this group will be greater than if the office be placed at the telephonic center. As the sum of all the parts can only be a minimum when each part is a minimum, it is only possible to secure this result by placing each office at the telephonic center of its own portion.

CONDITION II.

This condition requires that each division line between groups shall be perpendicular to, and bisect a line joining, the telephonic centers of the adjacent groups. Let Fig. 23 represent any territory whatever with any distribution of sub-stations. Suppose this territory to be sub-divided into two parts, the telephonic centers of which are at A and B, and let $A B$ be any straight line joining these centers. Also let $m_1 m_2 m_3$ be any sub-stations equidistant from $A B$. Let $a b$ be any division line drawn through the point midway between A and B. Then $A O + O m_1$ is equal to $B O + O m_1$ by construction and m_1 is equidistant from either A or B. Now, assume the line $a\,b$ to move in either direction to the points P or Q. Then

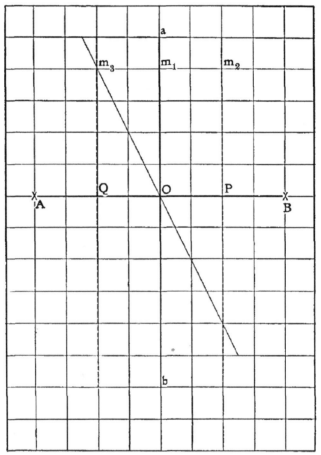

Fig. 23.—*Sub-division of Territory.*

$A\,O + O\,P + P\,m_2$ is greater than $B\,P + P\,m_2$ and $A\,Q + Q\,m_3$ is less than $B\,O + O\,Q + Q\,m_3$. Hence, any station to the right of $a\,b$ in its original position can

be carried to B with less mileage than to A, and any one
to the left can be carried to A with less mileage, also any
station on $a\,b$ can be carried indifferently to either office
with the same mileage; therefore, the division line must
bisect $A\,B$. Suppose now the line $a\,b$ to revolve about
the point O till it intersects any station m_3, which is at the
same distance from $A\,B$ as the station m_1. Since m_3 is on
the dividing line it is immaterial to which office it is
carried, for $m_3\,O + O\,A = m_3\,O + O\,B$. But either m_3
$O + O\,A$ or $m_3\,O + O\,B$ is greater than $m_1\,O + O\,A$
or $m_1\,O + O\,B$, hence the only line which passes through
O, all of whose points are at a minimum distance from
A and B, is one which is perpendicular to $A\,B$.

This demonstration has not recognized the condition
that all stations must proceed, rectangularly to the co-ordinate axes, to their respective offices. So *Condition II.*
must be modified to read that all points on the lines of
division between adjacent districts must be *rectangularly*
equidistant from the adjacent offices. For, in Fig. 24, let
A and B be the telephonic centers of two divisions of any
district, the lines of the paper standing for the street directions. Draw $A\,B$ according to *Condition II.*, the division
lines separating the two parts of the territory must bisect
and be perpendicular to $A\,B$ at O. Stations, however,
cannot reach A or B, save by routes that are parallel to
$A\,D$ or $A\,C$. Therefore, the shortest available distance
connecting the offices is $A\,C\,B$ or $A\,D\,B$. To determine
the division line, therefore, it is necessary to bisect $A\,C\,B$,
obtaining the point H, and $A\,D\,B$, obtaining the point I,
and to erect the perpendicular $H\,G$ and $I\,K$. By inspection it is evident that all points on $G\,H\,O\,I\,K$ are equidis-

tant rectangularly from either *A* or *B*. Hence this line will be the true division line separating the two parts of the district. All stations on the line may with the same

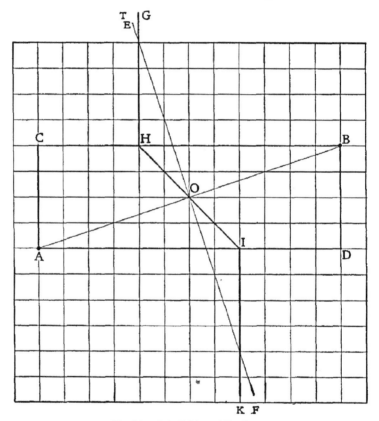

Fig. 24.—Sub-division of Territory.

mileage go to either office, while all stations to the left must proceed to *A*, and those on the right to *B*.

This condition of rectangular measurements leads to a special case when the telephonic centers are so located that the separating line joining them is at 45° with the co-ordinate axes. Refer to Fig. 25, in which *A* and *B* are the respective telephonic centers, and *A B*, the line joining them, is 45° to the lines of the section paper representing the street directions. Then *f A e* and *f B e* are

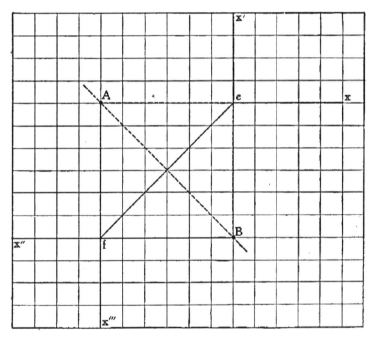

Fig. 25.—*Sub-division of Territory.*

the shortest rectangular routes between the two office locations. These lines must be bisected to determine the direction of the division lines, and the bisecting points will lie at the points *e* and *f*, the diagonal corners of the quadrilateral *f A e B*. At the points *e* and *f* draw the perpendicular *e x e x′ f x″* and *f‴*, including the spaces

$x\,e\,x'$ and $x''f\,x'''$. From inspection, all the stations within these areas can be taken indifferently, to either A or B, while the stations in the space $x''f\,A\,e\,x'$ must go to A, and those in $x\,c\,B\,f\,x'''$ to B.

Ordinarily speaking, compliance with *Conditions I.* and *II.* is sufficient to sub-divide any territory as may be de-

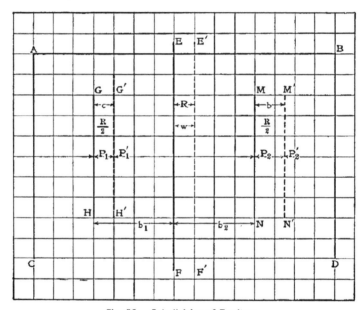

Fig. 26.—Sub-division of Territory.

sired, but it is conceivable to imagine such a distribution of stations as would permit the fulfillment of these require-ments when the position of the division lines is indetermi-nate, or could exist under the specifications given in more than one location. Therefore, for strict accuracy, some criterion is needed whereby it is possible to discover, if desired, which of several possible locations of telephonic

centers and corresponding division lines that all fulfill
the preceding conditions, will yield the desired minimum
mileage.

In Fig. 26 let *A B C D* represent any territory with
any station distribution, the lines of the paper representing the street directions. For simplicity assume the district to be divided into two parts, of which the telephonic
centers are P_1 and P_2, and let *E F* be the division line
between them, according to *Conditions I.* and *II.* Suppose some other location of this (at say *E' F''*) would also
carry out the conditions. As the line *E F* moves to its
new location *E' F'*, it will cross a certain number of stations, which may be denoted by *R*; so when it has arrived
at *E' F'* *R* stations will have been transferred from the
office at P_2 to that at P_1. But the law of telephonic center location requires an equal number of stations to be on
each side of a line drawn through the center parallel to
the co-ordinate axes, and P_1 and P_2 were so located.

If, now, *R* stations have been taken away from P_2 and
added to P_1, this equality is destroyed and P_2 and P_1
must be moved in the same direction as *E F* till each has
crossed $\frac{R}{2}$ stations. To accomplish this, suppose P_1 to
move *C* units into P_1' and P_2 to move *b* units to P_2'.
Let *G' H'* and *M' N'* be the respective new axes of *Y*
through the new telephonic center at P_1' and P_2'. Let
X_1 and X_2 be, respectively, the average abscissæ of $\frac{R}{2}$
stations from *G H* and *M N*, and *m* the average abscissæ
of *R* stations from the line *E F*. Let l_1 and l_2 be the distances of *E F* from P_1 and P_2 respectively. Let S_1 and

S_2 be the number of stations in office P_1, and office P_2 and L_1 and L_2 the average abscissæ in each division. Then $S_1 L_1$ and $S_2 L_2$ will, respectively, be the sum of all abscissæ in each division.

When $G H$ moves to $G' H'$ all stations on the left of $G H$ have their lines increased by an amount C. Those on the right of $G H$ have their lines decreased by C, while $\frac{R}{2}$ stations crossed by the moving line have their abscissæ changed by an amount $C - 2 x$. Similarly it is easy to show the change in abscissæ in office P_2. As the line $E F$ moves to $E' F' R$, stations are changed from P_2 to P_1. Originally, these stations had abscissæ represented by $R (l_2 - m)$. When transferred to office P_1' their abscissæ will be $R (l + m - C)$. It is now easy to calculate the sum of the abscissæ in the two parts of the given territory, both before and after the division line is shifted. The sum of the abscissæ *before* the line moves is $S_1 L_1 + S_2 L_2$. The sum afterwards is $S_1 L_1 + S_2 L_2 - R x_1 + 2 R m - R x_2$. Subtracting the latter expression from the former there remains the quantity $2 R m - [R x_1 + R x_2]$. This is the difference in mileage caused by the movement of $E F$ to $E' F'$.

Considering this expression, so long as $2 R m$ is greater than $R x_1 + R x_2$, it has a real positive value indicating that there is an increase in mileage by the movement of $E F$, and that the original location of this line is the preferable one. When $2 R m = R x_1 + R x_2$ the expression reduces to zero, showing that the new location has made no change in the value of the sub-station abscissæ, while if $2 R m$ is less than $R x_1 + R x_2$, the expression has a negative value, proving that in the new position the sub-station distances are decreased, and the arrangement of groups improved. $2 R m$ represents twice the sum of the distances of the sub-stations crossed by the moving division line measured from its original position. $R x_1$ and $R x_2$ represent the sum of the distances of the sub-stations crossed by the axes of Y in each case as they move from P_1 to P_1', and from P_2 to P_2', also measured from the original location. The criterion is, therefore, deduced that when twice the distance from the original position of the division line to sub-stations crossed by it as it moves, is greater than the sum of the corresponding distances of the sub-stations crossed by the co-ordinates through the several offices as they move, a change in the location of the division line is detrimental.

If the most complete and exhaustive analysis be desired, the expression $R\left[2m - (x_1 + x_2)\right]$ may be used to determine the effect of moving a division line over an entire territory, thus ascertaining the effect on sub-station lineage of every possible position of the line. If for successive points the value of this expression be plotted, a curve is obtained, giving the relation between sub-station lineage and location of divisions between offices. It is conceivable that several minima could, under some circumstances, be found from which the best one could, by inspection of the curve, be discovered.

By applying the same process to the other co-ordinate axis, and combining the results of both analyses, the effect of every possible sub-division may be plotted. At first sight the process appears formidably laborious, and there is no gainsaying that to secure valuable results considerable time must be expended, yet with a good map, a square arranged to slide rectangularly over it, whose edges are graduated to read directly in wire mileage. and an adding machine, it is surprising how fast a skillful operator will sub-divide a district. Again, exchange sub-division is chiefly of value to large and telephonically very populous territories, where the saving of a very small wire mileage on each of the lines of many thousand stations may aggregate a very handsome return on a few weeks' labor in the drawing-room.

Practically the process is much simpler than the theoretical description would indicate, for provided with a good map on which stations are plotted, and armed with a fair local knowledge of the various business interests of the territory the skillful telephonist will, by inspection of the map, pick out the approximate groups into which the whole territory should be divided in a general way. The telephonic center of the subordinate groups may then be quickly located and the boundary lines sketched in accordance with *Conditions I.* and *II.*, and the wire mileage to all stations calculated. For a large territory it is usually best, for reasons to be subsequently recited, to make several sub-division designs, and compare the several wire plant expenses with each other; therefore, for any city of, say, 500,000 inhabitants, it is well to make four or five trial sub-divisions, splitting the territory into, for example, four, six, eight, and ten groups, and plotting a resulting curve of the relation of sub-station mileage.

Every time that a subordinate group in any given territory is subdivided the process of division slightly reacts on all other groups, so for strict accuracy, it is necessary after

any group is split, to correct the boundary lines and telephonic centers of all others, but practically this error is too
small to be entertained. To illustrate the process of finding the curve of sub-station mileage and number of offices,
together with the effect of probable error when the mutual

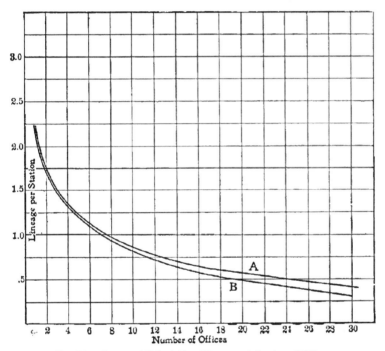

Fig. 27.—Curves of Sub-station Mileage and Number of Offices.

adjustment of boundary lines and telephonic centers
referred to is neglected, turn to Fig. 27.

The curve *B* gives the relation between the number of
offices and average length of sub-station lines in a hypothetical territory resembling an average city of 500,000 in-

habitants.　The number of offices varies from one with an
average sub-station mileage of 2.25 to 30 with an average
mileage of .30.　In this case all the corrections have been
made.　Curve *A* is plotted without adjusting adjacent
boundary line and telephonic centers when a sub-division
is made.　The failure to make this adjustment returns results that are somewhat too high.　For four offices the un-
adjusted curve is one-twentieth of a mile, or about 3.75 per
cent in error.　But for 30 offices the error is about 27 per
cent.　But as it is not possible to give a general error curve
applicable to all territories, it is best to use the uncorrected
mileages for the various trial sub-divisions, and when the
actual number of offices is finally selected to carefully ad-
just all boundary lines and telephonic centers to each other.

CHAPTER X.

THE RELATION BETWEEN THE NUMBER OF OFFICES AND THE MILEAGE OF THE TRUNK LINE PLANT.

WHEN any territory is served by a single office, each sub-station may be immediately placed in talking relations with any other one in the whole group by the simple process of interconnecting the two sub-station lines by a flexible cord; for on the single office multiple switchboard system, every subscriber's line is within the reach of any operator. But when a given territory is sub-divided, and has more than one office, it is no longer possible to interconnect all sub-stations, for the lines of some will converge to, and terminate in, one of the offices, while others will end in some other office. It is therefore necessary to provide some means whereby stations tributary to one office that desire communication with those ending in any other office, may be accommodated. This is accomplished by providing between all of the various offices taken in pairs a set of special wires called *trunk lines*.

It has been shown that the total sub-station mileage for any territory decreases rapidly as the group is sub-divided into smaller and smaller portions; but it is evident that this very endeavor to minimize sub-station mileage will be accomplished with a multiplication of trunk lines, for with each additional sub-division another office is added, from which, to all other offices, trunk lines must be provided. Further, the smaller the portion of any territory that is served by any one office, the larger will be the portion of the exchange foreign to it, and the greater will be the probability that sub-stations will desire to talk outside, and not inside, of the boundaries of the office in which they terminate. So as sub-station mileage is decreased, by subdividing any territory, trunk-line mileage is increased, and it is easily imaginable that the process of sub-division could be carried to such an excess as to cause the increments of trunk mileage to overbalance and exceed the decrements of sub-station mileage. Evidently the true minimum total wire mileage in any case is secured by making the sum of all sub-station mileage, plus that of all trunk-line mileage, a minimum, and the method of discovering this condition must now be described.

. The total trunk-line mileage in any territory will be proportional to the number of trunk lines needed, and to their length. The number of lines required will depend upon the volume of business to be carried, and a factor that might be termed the *Specific Trunking Capacity*, or, in other words, the number of messages per unit of time that it is practical for a trunk line to transmit. The volume of business will depend upon the number of sub-stations; the number of messages originated per unit of time, per station, and the ratio of the number of messages sent outside of the office under consideration to the total number originated. The length of the trunk lines will depend on the size of the exchange, and the number of offices into which it is split. Let S represent the number of sub-stations in any exchange, and T the average number of originating messages per day, then ST will be the total number of messages originated; this quantity is usually called the *Originating Traffic*. Let S' and T', and S''' and T'', etc., represent respectively the number of sub-stations, and their daily average originating messages, in the various offices into which the exchange may be sub-divided, then $S'T'$ and $S''T''$, etc., will denote the originating traffic of each of the individual offices. Let R' and $R,''$ etc., represent the ratio of the messages sent out of offices S', S'', etc., to their originating traffics, then $R'S'T'$ and $R''S''T''$, etc., will denote the daily business sent out of the offices S', S'', etc. For brevity it is convenient to call this quantity the *Out Trunk Traffic*. If Q stands for the Specific Trunking capacity per line per day,

$$\frac{R'S'T'}{Q}; \quad \frac{R''S''T''}{Q}, \text{ etc.,}$$

will be the number of trunk lines needed for the offices S', S'', to all the rest of the exchange and

$$\frac{R'T'}{Q}; \quad \frac{R''T''}{Q}, \text{ etc.,}$$

will be the number of trunk lines per sub-station in the several offices. For brevity call these quantities respectively the *Office Trunkage and the Subscribers' Trunkage*. If L' and L'', etc., denote the average length in miles of the trunk lines from the offices S', S'', etc., to all other offices, then

$$\frac{R'S'T'L'}{Q}; \quad \frac{R''S''T''L''}{Q}, \text{ etc.,}$$

is the trunk-line mileage from offices S', S'', etc., to all other offices; call this quantity the "*Trunk lineage*," then

$$\frac{R'T'L'}{Q} \text{ and } \frac{R''T''L''}{Q}, \text{ etc.,}$$

is the *Trunk lineage per sub-station.* For every exchange the preceding quantities $S\,T$, $S'\,T''$, etc., $R\,Q$ and L will have certain specific and definite values which must be ascertained and used in order to secure the best design. But average values derived from extended experience will best exemplify the method of calculation, and will yield probable average results, that may be taken as general guides.

The values of S and T will vary with the size of the city under consideration. From present indications it seems quite certain that cities of 500,000 population and over will soon have at least 40,000 to 50,000 subscribers. Those of 200,000 to 300,000 inhabitants from 20,000 to 30,000 telephones, while towns of 75,000 to 100,000 people should be credited with not less than 10,000 stations. It is difficult and somewhat hazardous to assign a value to T on account of the rapidly changing character of the telephone business. As an exchange grows in size the increasing number of subscribers give greater and greater opportunities for conversation, and if all parties equally availed themselves of this possibility, the traffic would vary with the square of the number of sub-stations. But this is never the case; on the contrary, increase in traffic rarely keeps pace with the increase in sub-stations. The depression in the traffic rate is partly due to the fact that people having the greatest amount of telephone business are the first ones to subscribe, while those less telephonically busy come in afterward; and partly owing to the recent introduction of the coin box, slot machine, and nickel telephone, or other forms of measured service which are rapidly creating a very large telephonic population who wish to " pay as they go," and are strongly averse to the contract and flat rate system. With the flat rate method there is no incentive to any economy in traffic; and the subscriber, his friends, acquaintances, and employes use the instrument *ad libitum.* But with any form of measured service, where the rental is at least in some degree proportional to the traffic, there is a strong motive for economy in the use of the instrument, a most powerful factor to depress the call rate. Taking all things into consideration, it seems probable that even the busy offices in the center of the largest cities will not average over twenty originating messages per day, while those in the residence districts, or on the outskirts, should be credited with ten. For offices similarly located in medium sized cities, a message rate respectively of sixteen and eight will be assumed, while for towns from 75,000 to 100,000, ten messages per station per day seems fair.

As any district is repeatedly sub-divided into a constantly increasing number of offices, the size of each office and the number of sub-stations it contains constantly diminishes. Therefore S, or the number of sub-stations, etc., is the independent variable in both the trunk-line and sub-station line problem, and will, for this part of the discussion, be assumed to vary from 500 to 7000, as it seems that smaller offices than 500 would never be considered, and that the economy of those larger than 7000 is certainly doubtful, as there is no American experience with central stations even of this magnitude.

The value of R represents the proportion of originating messages that any office will send beyond its boundary lines. If it should be assumed that each ·subscriber in all of the various offices into which any territory may be divided originates an equal number of messages, and further, that each subscriber transacts an equal amount of business with every other subscriber, the messages trunked out of any one of the separate offices will bear the same ratio to the total messages originated by all the subscribers in the office under consideration, that the sum of the subscribers in the given office bears to the sum of all the subscribers in all the offices. So if S be the total number of sub-stations in the entire territory, and S' the number in any particular office, then under the preceding hypotheses, the percentage of messages trunked out of the particular office will be

$$\frac{S - S'}{S}.$$

This is a ratio; but if T' be the average number of origi̇nating messages in the office in question, $S' \, T'$ will be the total number of messages originated by the office, and then

$$\frac{(S - S') \, S' \, T'}{S}$$

will be the out-trunk traffic. This formula assumes an equality in traffic over the whole exchange; but if the average traffic T' of the office S' differs from the average traffic T of the whole territory, this discrepancy must be recognized by introducing the factor T' as follows:

$$\frac{ST - S'T'}{ST} \, S'T', \quad \text{or} \quad \frac{SS'TT' - S'^2 T'^2}{ST},$$

which gives the out-trunk traffic of the office under consideration.

The preceding formula is based on the assumption that each person will always send an equal number of messages to all others. Such an hypothesis could only be true in a case where every individual was acquainted with, and equally interested in, every other one in the whole community. In small villages the condition of equality of traffic is sometimes approximately realized, for in such cases every one is apt to know, and likely to have some dealing with, all others. But in cities, particularly the larger ones, groups are formed, each one gathering around him a particular circle of friends, acquaintances, and business associates, and ignoring all others, thus the telephonic business of each individual flows in certain well-defined channels from which it rarely and only spasmodically departs. But owing to the complexness of society there is a constant overlapping of the acquaintance circles. Thus A and B are mutual acquaintances, and each knows a hundred other people. Of the hundred in A's circle B is acquainted with fifty, and of the hundred in B's circle A knows fifty. A and B talk together; B talks to fifty people outside A's circle, and A talks to fifty outside of B's circle; each talks to a hundred mutual acquaintances, and neither A nor B sends any amount of business to the great mass of sub-stations outside the limits of their particular coteries. To recognize this condition it is necessary to introduce in the preceding expression a coefficient that may be denominated the "*Acquaintance factor.*" If q

represents this factor, the actual out-trunk traffic will be ascertained by the formula,

$$\frac{q\,(S\,T - S'\,T')\,S'\,T''}{S\,T}.$$

The value of q varies between quite wide limits. In cases where there are two or three offices of nearly equal size and serving similar districts, about 68 per cent will hold true. When there are several small offices on the outskirts working into larger ones near the center, 85 per cent to 95 per cent will represent q for the smaller offices, and 60 to 65 per cent for the large ones. In the cases of a number of medium-sized offices, 70 per cent to 80 per cent is a fair average; while in the extreme of one large office and one small one 40 per cent to 50 per cent will indicate the factor for the large one, and 92 per cent to 98 per cent for the small one.

The preceding expression will give such a value for the trunk traffic as would exist if the incoming and outgoing messages were always numerically equal in each office, and here a misconception often arises. Take, for example, an exchange of two offices, — one, A, in the center, and the other, B, in the outskirts of a large town. Suppose A sends B ten messages, and B sends A one hundred; there will be a total of one hundred and ten outgoing trunk messages and one hundred and ten incoming trunk messages; but the incoming trunk messages *in* A will be ten times as great as the incoming messages *in* B. In the formula the only known quantities are the number of stations in the whole exchange and average message rate; also the number of stations and average call-rate in the particular office under consideration. Therefore the formula gives the mean value between the numbers representing the incoming and outgoing trunk messages. Usually the difference is very slight, not more than three or four per cent, and it is only in case of a very large office and a very small one that any practical difference would exist. If there *is* a sensible difference between the incoming and outgoing messages, it is possible to more closely approximate to the true value by adding or subtracting from the result given by the formula such a quantity as will recognize this difference. The total number of originating messages is the only known quantity, hence the desired correction must be expressed as a percentage thereof. Let p be the percentage, then $\dfrac{p(S'T')}{100}$ will be a quantity either to be added to or subtracted from the mean value in order to ascertain the desired true number of out-trunk messages. The correction must be additive in cases where the outgoing work is in excess of the average, and subtractive where it is deficient.

The complete expression for the out-trunk traffic is therefore:
$$\frac{q(ST - S'T')\,S'T'}{ST} \pm \frac{p(S'T')}{100};$$
but usually sufficient accuracy is obtained by using the abbreviated form,
$$\frac{q(S - S')}{S}\,S'T';$$
in which q has a value ranging from .65 to .85, depending upon the relative sizes of the offices S and S', and from these expressions the following values for R are immediately deducted:
$$R = \frac{q(ST - S'T')}{ST} \text{ or } R = \frac{q(S - S')}{S}.$$

CHAPTER XI.

SPECIFIC TRUNKING CAPACITY.

To determine the specific trunking capacity Q it is necessary to consider in some detail how trunk lines are handled, in order that the limitations surrounding them may be given full weight. When a subscriber calls the central office in which his line ends, the act of signaling displays an indicator (usually a drop or small incandescent lamp) placed in sight of an operator (termed a subscriber's or A operator) before whom the line of the particular subscriber terminates. On perceiving the signal the A operator inserts a plug into the answering jack of the subscriber calling, and switching her telephone set into communication with the cord attached to the plug placed in the jack gives the familiar request, "Number, please?" The subscriber replies, stating the number of the party desired, together with the name or designation of the office in which the called party is to be found; for in all multi-office exchanges each office has its own special appellation. Learning that the called subscriber is in a foreign office the A operator must obtain a trunk line leading to the office desired and gain the attention of the operator (usually called a B, or incoming trunk operator) before whom this trunk line appears, in order that this second operator may call the wished-for party, and connect his line to the trunk line. When this is done the A operator in the originating office can establish communication by inserting into the trunk-line jack the companion plug to the one already placed in the jack of the calling subscriber. There are two methods in common use whereby the A operator may gain the attention of the B operator. The first and oldest method requires the use of a signal for each trunk line placed in sight of the B operator, capable of being actuated by the A operator. Under this arrangement each A operator must either have a separate and independent set of trunks to each B operator, which necessitates an enormous trunk-line plant, or if the same trunk lines appear, or are multiplied in front of a number of A operators, some means must be provided to inform the several operators which lines are in use, for if this is not done, two operators may plug different parties on the same trunk line, and the consequent *double connection* results in "confusion worse confounded" to the exasperation of the subscribers and the humiliation of the operators. It is easy to devise an automatic signal associated with each trunk line placed before the A operators that shall inform them whenever a line is engaged. But such signals take up considerable space in the face of the switchboard, which cannot be spared in these days of five thousand line switchboards, wherein every sixteenth of an inch must be gained to make room for the multiple jacks. So it is usual to arrange that trunk lines in use may give the familiar "*busy test,*" or emit a sharp click in the A operator's receiver when the jack rings with which they are associated are touched with the tip of the connecting plug.

In the case of large and busy offices many trunk lines are needed, and during the rush hours of the day an operator may have to test a dozen or twenty lines before finding one that is disengaged. This takes considerable time at a period when it can be ill afforded; also the operator is quite liable to error, and in spite of the busy test may take engaged lines, all of which is exceedingly injurious to the service. But omitting errors from consideration the course of business is as follows: On receiving the subscriber's order the *A* operator tests the trunks leading to the desired office; selects one disengaged; inserts the connecting plug, signals on the trunk line, by ringing or otherwise. On perceiving the trunk-line signal the *B* op-. erator restores the signal to its normal position, inserts an answering plug into the trunk-line jack in the second office, switches her telephone set onto the cord, and asks the *A* operator for "Number." Then the *A* operator gives the *B* operator the order, on receiving which the *B* operator places the connecting plug into the jack of the called subscriber and rings.

To avoid the loss of time, errors, and other imperfections of the *Signal Trunk System* the *Reverse Call Circuit Method* has been devised, and has practically supplanted the old plan in all but small offices. In this method one or more special pairs of wires are provided to each *B* operator's telephone to which the various *A* operators can connect their transmitter by the simple function of pressing a button, or *call circuit key*. The trunks extend from jacks in front of the *A* operators and end in cords and plugs before the *B* operators. The course of business is as follows: When an *A* operator receives an order from a foreign office she touches the proper call circuit button, and speaks the desired number directly into the ear of the proper *B* operator. The *B* operator picks up a plug attached to a disengaged trunk, inserts it into the jack of the desired party and rings, simultaneously pronouncing *back* over the call circuit to the *A* operator the number of the *trunk line selected*, and the *A* operator inserts the connecting plug into the jack of the trunk line thus designated. The simplicity, speed, and accuracy of this plan leave little to be desired for the transaction of trunk line business.

If the average length of time that a trunk line is in use during the transmission of messages be known, its practical specific trunking capacity can at once be calculated. The average time of occupancy must not only include the interval of time that parties are actually engaged in conversation, but must allow for the time taken by the *A* operator to obtain the trunk; for the *B* operator to find the jack of the called subscriber, and signal; for the called subscriber to answer the ring of his telephone bell; and finally, after conversation is completed, a sufficient interval to enable both operators to restore the trunk line to its normal condition, ready for the next message. Of all these various intervals, the time occupied by the subscriber in answering the telephone bell is the most uncertain factor. In the business hours of large cities answering is usually speedy and reliable, for business men are trained to prompt, quick action; and further, usually have a pretty fair idea of the inconvenience to which others are subjected by delay in replying, and the unjustness of such a course. Again, one or more employes are usually assigned the duty of responding to telephone messages, and general office discipline ensures immediate attention. Small city stores, and country town business houses come next on the list, with greatly extended and very variable times of answering, while residences, particularly those in small towns and villages, are telephonic terrors to their exchanges, in the utter irresponsibility displayed to tele-

TABLE No. IV.

Analysis of Time occupied by Trunk Line Messages.

1	PROBABLE TIME.		
	2 MINIMUM.	3 AVERAGE.	4 MAXIMUM.
A. Signal trunks			
A operator answers . . .	2.0 sec.	5.0 sec.	7.0 sec.
B operator rings	5.0 sec.	12.0 sec.	20.0 sec.
Subscriber answers . . .	10.0 sec.	15.0 sec.	30.0 sec.
Conversation	10.0 sec.	120.0 sec.	300.0 sec.
Disconnect	3.0 sec.	6.0 sec.	10.0 sec.
Total	30.0 sec.	158.0 sec.	367.0 sec.
B. Reverse call circuit trunks . .			
A operator answers . . .	2.0 sec.	3.0 sec.	5.0 sec.
B operator rings	3.0 sec.	5.0 sec.	10.0 sec.
Subscriber answers . . .	10.0 sec.	15.0 sec.	30.0 sec.
Conversation	10.0 sec.	120.0 sec.	300.0 sec.
Disconnect	2.0 sec.	3.0 sec.	6.0 sec.
Total	27.0 sec.	146.0 sec.	351.0 sec.

phone calls. In a case in mind (an actual fact) a worthy housewife finished a loaf of bread she was kneading, when the telephone bell rang, and a considerable fraction of an hour later when she answered, soundly berated the unfortunate operator because the calling party had not been compelled to await the completion of the breadmaking.

The length of time occupied in conversation necessarily
varies, between extremely wide limits, owing to the varying
nature of business to be transacted; and while instances
are on record of lines being held for an hour and a half,
such are the exceptions that only emphasize statistics
showing that scarcely five per cent of all messages last
over three minutes, and some seventy per cent are completed
in less than a minute and a half. City messages are
shorter than country calls, and business communications
much more terse and abrupt than those from residences.
The times needed by the operators to find the trunk line,
signal the called subscriber, and disconnect when conversation is completed are fairly regular and easily deter-
minable by observation. In the foregoing Table No. IV.
the consumption of time for the various avocations indicated is given.

Columns 2, 3, and 4 of this table exhibit the probable
consumption of time in the performance of the functions
specified in column 1. The second column is the minimum
time achieved with the very best service, column 3 the
average time ordinarily to be counted on, while column 4
the maximum allowable time compatible with service even
remotely satisfactory. From these statistics it would
appear that on an average a trunk line is actually engaged
a little less than three minutes per message, and that with
an allowance of thirty seconds for lost time a specific
trunking capacity of 20 messages per line per hour might
be expected. Occasionally, under exceptionally good
management, this amount of work is realized, or even
slightly exceeded, but on the average fifteen calls per line
per hour is all that can be calculated on, and even this

quantity can only be handled in large and busy exchanges, under the most vigilant superintendence. The reason for so small a specific trunking capacity is not far to seek. In no business is the load factor so low as in telephony. Take, for example, the average subscriber's line, and assuming the average message to consume two minutes the line has a possible daily load factor of 720 messages. But even in the busiest offices of the largest city it is rare to find an average call rate of over thirty messages per station per day, or a load factor of only 4.15 per cent. For fourteen hours of the twenty-four most lines are absolutely idle, for few but residence telephones call outside of business hours. For trunk lines the load factor is somewhat better. From the preceding table it is seen that a trunk line would have a possible daily capacity of, say, 540 messages for signal trunks and 580 messages for call circuit systems. In well managed busy offices trunk lines will carry 150 messages per day, but taking large and small offices together, and allowing for inefficiency in management, it is doubtful if the trunk-line plants of all the exchanges average a hundred messages per line per day. Thus the load factor might, under favorable circumstances, reach 26 to 27 per cent, but would only average from 15 to 18 per cent.

A telephone plant also differs markedly from most other electrical installations in its absolute inability to respond to any forcing. A dynamo will answer to a fifty per cent overload for a considerable period without injury, but a trunk line can by no possibility carry out *one message at once*. It is either working at *no* load or a *full* load; no other condition is, at present, conceivable. It is equally

futile to press the operators, for the present rates of work
are so taxing that almost the slightest excess so increases
the nervous strain that the operators lose presence of mind,
become confused and break down. There is also no mo-
mentum, no inertia to the parts of a telephone plant, and
no possibility of any storage device that will enable even
momentarily a rush of business to be carried over. When
a subscriber calls he demands immediate attention and
complains bitterly at the delay of a second in the response
of the operator. To be told that a party in a foreign exchange could not be obtained because all the trunk lines
were busy would invite an unquenchable storm of criticism
from the modern telephone subscriber. So it is necessary
to supply a trunk-line plant so large as to be able to carry
almost any conceivable overload. Not only must there be
ample facilities to carry the peaks and rushes of the busy
hours in the daily load, but the exigencies of holiday traffic,
coupled with the extra pressure of bad weather, must be
reckoned with, and provided for.

As the trunk-line plant usually consists of cables buried
in the ground it is not possible to put it in and out of service at pleasure, but it must be designed and installed a
long time in advance of the necessity of its use, and once
in place must remain till its term of service is over.
There is one aspect, however, in which the trunk plant
shows some elasticity. In cases where there are three or
more offices in an exchange, it is always possible to connect any office indirectly with any other office by looping
together in a roundabout fashion the lines through two or
more offices, when the direct trunks between the pair of
offices are exhausted or incapacitated. So, for the preced-

ing reasons it is unsafe to assume a trunk-line load Q, of more than 150 messages per day from large and carefully managed offices, 100 messages per day from medium-sized offices, and 80 messages per day from small offices.

In a preceding paragraph it was shown that the office trunkage symbolized by T was given by the formula $R\,S_2'\,T'$. Substituting in this expression the values of R and Q just deduced, and denoting with small numerals the limiting values of the literal factors, as is common with the sign for integration, the actual number of pairs of wires needed to carry the out-trunk traffic from any office to the rest of the exchange is obtained as follows:

$$\text{Number of pairs} = \frac{q^{.85}}{Q^{150}_{80}}\left[\frac{ST - S'T'}{ST} \times S'T' \pm \frac{P S'T'}{100}\right];$$

or approximately,

$$\text{Number of pairs} = \frac{q^{.85}}{Q^{150}_{85}}\left(\frac{S - S'}{S}\,S'\,T'\right).$$

This is the actual number of lines needed to carry the out-trunk traffic, but experience has demonstrated the wisdom of installing a slight excess in trunk-line plant over and above the calculated number needed, that may act as a factor of safety, to care for unexpected fluctuations in traffic, and provide for possible growth.

If n represent the excess pairs advisable to thus install for each office, and N the number of offices in the exchange, then $n\,(N-1)$ is the total excess trunkage. The indications of practice are that in small offices where the load fluctuations are likely to be severe, and there is a probability of rapid growth, n should have a value of from 3 to 5, while in large offices with many lines, steadier loading, and slower growth, it may be as low as .5. When signal trunks are employed the true trunkage T will be the sum of the actual pairs needed plus the excess pairs, or,

$$T = \frac{q^{.85}_{.65}}{Q^{150}_{80}}\left[\frac{ST - S'T'}{ST} \times S'T' \pm \frac{P\,S'T'}{100}\right] + N^{5}_{.5}(N-1),$$

or approximately,

$$T = \frac{q^{.85}_{.65}}{Q^{150}_{80}}\left(\frac{S - S'}{S}S'T'\right) + N^{5}_{.5}(N-1).$$

In these expressions there is no allowance for the necessary pairs of wires to serve for call circuits to the B operators when the reverse call circuit system is employed. Evidently there must be at least one call circuit from the given office to each of the others, so there will be at least $N - 1$ call circuits, from the office aspect, and possibly more if the volume of out-trunk traffic demands. With the call circuit system good practice shows that a B operator can handle reasonably 3000 messages per day. Therefore the out-trunk traffic divided by 3000 indicates the number of order wires needed from purely a *traffic* standpoint. Using the expression previously deduced for the out-trunk traffic, and dividing the same by 3000, the number of pairs needed for call circuit traffic is developed as follows:

Call circuit pairs $= q_{.65}^{.85}\left(\dfrac{S\,T - S'T'}{3,000\ S\,T}S'T' \pm \dfrac{P\,(S'\,T')}{300,000}\right);$

or approximately,

Call circuit pairs $= q_{.65}^{.85}\left(\dfrac{S - S'}{3,000\ S}\ S'\,T'\right).$

Now if the above result is *less* than $N - 1$, then $N - 1$ pairs must be employed, or if it be greater, then the value of the formula must be taken; algebraically this is stated as follows:

Call circuit pairs $=$

$$N - 1 + q_{.65}^{.85}\left(\frac{ST - S'T'}{3,000\ ST}\ S'T' \pm \frac{P(S'T')}{300,000}\right) - [N-1],$$

or approximately,

Call circuit pairs $= N - 1 + q_{.65}^{.85}\left(\dfrac{S - S'}{3,000\ S}\ S'\,T'\right) - [N-1],$

the quantity in the brackets being taken only in its arithmetical sense as a positive quantity. The complete expression for the trunkage from any office to the whole of the rest of the exchange thus becomes for reverse call circuit system:

$$T = \frac{q_{.65}^{.85}}{Q_{80}^{150}}\left[\frac{ST - S'T'}{ST}\ S'T' \pm \frac{P(S'T')}{100} + N_{.5}^{5}(N-1)\right]$$

$$+ N - 1 + q_{.65}^{.85}\left(\frac{ST - S'T'}{3,000\ ST}S'T' \pm \frac{PS'T'}{300,000}\right) - N - 1,$$

or approximately,

$$T = \frac{q^{.85}_{.65}}{Q^{150}_{80}}\left(\frac{S-S'}{S}S'T' + N^{5}_{.5}(N-1)\right) + \left[N - 1 + q^{.85}_{.65}\right.$$

$$\left.\left(\frac{S-S'}{3,000\,S}S'T'\right) - (N-1)\right].$$

To ascertain the complete trunkage for the entire territory, the preceding expression must be applied to each office in the exchange successively, thus obtaining the trunkage for each one, and finally the summation of all these separate results will give the desired total trunkage.

In order to complete the design for an interlacing trunk system it is necessary to know the number of trunks and call circuits from each office to every other office, the offices thus being taken in pairs. The preceding formula can be used for this purpose, only in this case S is the sum of the sub-stations in the two offices under consideration, while S' is the stations in the office *from* which the trunk lines are to extend, T is the average originating traffic in both offices, and T' is the originating traffic in the office S'.

It is often urged that a complete system of this kind requires the expenditure of more wire plant than the business will justify, and that a saving could be made by resorting to indirect trunking, or the transmission of messages from one office through a second office to a third office. When the daily business is very small and the distance between offices great, the employment of signal trunks instead of the call circuit system will make a justifiable saving in wire plant. But experience has shown that the resort to indirect trunking is never an economy. The time consumed in sending trunk messages through three operators, and the concomitant blunders, injures the service far more than can be counterbalanced by the annual saving in cost of a direct trunk line in all ordinary cases. Signal trunks can, on a pinch, be designed to serve as both outgoing and incoming trunks in both offices, and it must be a very small office, in a very small exchange, that will not transact a hundred messages a day; an office of twenty-five lines should do more than this, and it is rare that so small an establishment would pay under any circumstances.

CHAPTER XII.

TRUNK LINEAGE.

KNOWING the number of trunk lines, the only missing factor necessary to determine the trunk lineage is the length of the trunks. It is impossible to predict the exact length and location of trunk lines in any exchange, until the specific territory shall be examined, districted, and the telephonic centers located. But it is feasible to ascertain for any group another quantity that is proportional to the trunk lineage needed with offices of varying areas, and which may be substituted for the true trunk lineage, passing under the same name, till such time as the telephone centers *are finally* located and the precise trunk lineage can be ascertained, and which will serve as a guide to determine the relative relations of sub-station lineage and trunk lineage when the territory is split up into varying numbers of offices.

It will be remembered that the average length of the sub-station lines is determined by dividing the sum of all the sub-stations mileage in any district by the total number of sub-stations, and that a series of curves was given showing the probable relation between the length of sub-station lines, and the number of subscribers tributary to offices of various sizes, in large and small cities. Now,

if any office is to deliver messages to sub-stations that
lie outside its boundaries, the average distance that such
messages travel will evidently be the sum of the distances,
measured along streets, from the office to all sub-stations
outside of the boundaries, divided by the number of stations.   The larger the territory embraced by the office
under consideration the greater will be the distance from
its telephonic center to sub-stations lying outside its
boundary, and the smaller will be the number of these
foreign sub-stations.   Hence the average distance that
messages must thus travel will increase with the size of
the offices. This quantity is called the *average trunking
distance*. The average sub-station mileage was ascertained
by plotting the stations on a map and scaling from the
various telephonic centers to which they were tributary.
In the same manner the average trunking distance may
be ascertained, by scaling on the map from the various
telephonic centers to all stations outside the boundaries
of the office. It is easy to see that while the average
trunking distance is not the same as the trunk lineage, or
rectangular distance between telephonic centers, it is proportional thereto. Now, if the trunkage for subscribers,
obtained by dividing the office trunkage by the number of
sub-stations, be multiplied by the average trunking distance, a quantity that represents the trunk lineage per
subscriber is obtained. If this calculation be made for
several offices of various size, it is easy to plot a curve of
the results, and compare the subscribers' trunk lineage in
offices of various sizes, and determine which gives the
best results.

In Figs. 28, 29, 30, 31, and 32 all of the preceding

quantities have been calculated and plotted for five typical cases, namely, Fig. 28 contains the data applicable to
offices located in the center of a large city, say, of 500,000
inhabitants; Fig. 29 the same curves for an office located
in the outskirts of the same city; Figs. 30 and 31 apply
to offices located respectively in the center and outskirts
of a medium-sized city of about 200,000 population, while
Fig. 32 is for the center of a small city or large town of,
say, 75,000 to 100,000 people. All five of the illustrations are plotted to the same scale and arranged in the
same manner, so that comparison is easy. There are nine
curves on each sheet, as follows:

Curve No. 1. Number of originating messages.

Curve No. 2.

Theoretical percentage of messages trunked $= \dfrac{S - S'}{S}$.

Curve No. 3.

Actual percentage of messages trunked $= \dfrac{q\,(S - S')}{S}$.

Curve No. 4. Out-trunk traffic $=$ Curve 1 $\times$ Curve 3.

Curve No. 5. Out-trunk traffic per station $= \dfrac{\text{Curve } 4}{S}$.

Curve No. 6.

Office trunkage $= \dfrac{\text{Curve } 4}{Q}$
$+$ reserve trunks and call circuits.

Curve No. 7. Sub-station trunkage $\dfrac{\text{Curve } 6}{S}$.

Curve No. 8.

Trunkage distance, ascertained from map measurement.

Curve No. 9. Sub-station trunk lineage $=$ Curve 7 $\times$ Curve 8.

On the bottom of each sheet the axis of X is the independent variable, and gives the size or number of stations

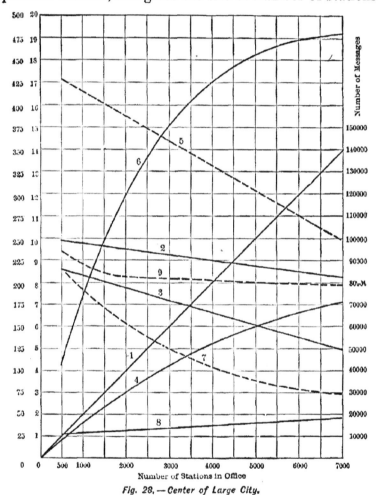

Fig. 28. — Center of Large City.

in the office under consideration. The vertical scale on the right hand indicates the number of messages, and is used with Curves 1 and 4. Thus in Fig. 28, relating to the center of a large city, the total number of stations in the whole exchange is taken at $S = 40,000$, the average originating messages T' and $T'' = 20$, Q is assumed to be 150, and q to vary, say, from .65 to .85, depending on the size of the office. Then in an office of, say, 4000 stations, the number of originating calls would be found by following a vertical line from 4000 on the lower horizontal scale till it intersects Curve 1, and then a horizontal to the right finding 80,000 on the scale of "Number of Messages." Similarly the out-trunk traffic for the office is obtained in the same manner, by following a vertical from 4000 to the intersection with Curve 4, thence a horizontal to the message scale finding 52,000. On the left hand of each sheet there are two scales of equal parts, the outer one reading from 0 to 500, and the inner from 0 to 20. The inner scale applies to Curves 2, 3, 5, and 8. Thus on Fig. 28, if it be desired to ascertain the theoretical percentage of messages trunked from an office of 2000 stations, follow a vertical from 2000 on the lower scale till Curve 2 is intersected, thence a horizontal to the inner scale on the left hand reading 95 per cent. The actual per cent of trunking is found on the same scale by stopping at Curve 3, and reading 87 per cent, while the out-trunk traffic per station is found by continuing till Curve 5 is intersected and then reading 15.4 messages on the same inner left-hand scale. Curve 8 exhibits the data pertaining to the average trunking distance, as determined by map measurements. Taking the various areas needed to include 1000, 2000 and 3000 stations, etc., scale measurements were made to all sub-stations beyond these boundaries, and the

sum of the distances (*via* streets) thus obtained divided by
the number of stations, this fixing the average trunking

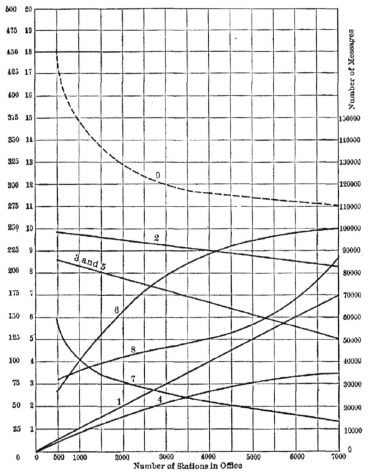

Fig. 29. — *Outskirts of Large City.*

distance.   In an office of 3000 stations the average trunking distance would be found by following a vertical from

3000 to the intersection with Curve 8, and thence a horizontal to the inner left-hand scale, finding 1.4 miles. The outer left-hand scale is devoted to Curves 6, 7, and 9. For Curve 6 (the total trunkage) this scale reads directly

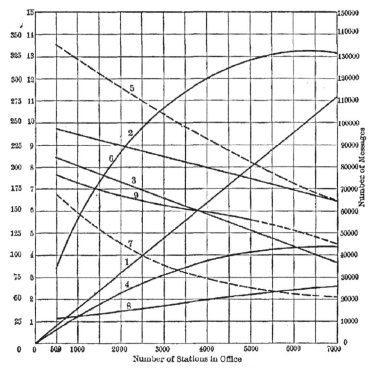

Fig. 30. — Center of Medium City.

in whole numbers; thus, to ascertain the total trunkage (including call circuits) that is needed from an office of 2000 stations follow a vertical line from 2000 on the lower scale to the intersection with Curve 6, thence a horizontal to the outer left-hand scale, finding 300. For Curves 7 and 9 a decimal point must be prefixed to all

the numbers of this scale. Thus the sub-station trunkage
and sub-station trunk lineage for an office of 3000 stations
are found by following a vertical from 3000 to Curves 7
and 9, and thence horizontals to the outer left-hand scale
to be respectively .1225 of a trunk line per station, and
.2025 miles per station. The curves of Fig. 29 are
plotted on the assumption that the office is in the outskirts of the same city, and therefore works *into* an exchange of 40,000 stations with an average originating
message rate of 20. The originating message rate for
the office is set at 10; the values of Q, q, and other factors
are the same as for Fig. 28, while the average trunking
distance is compiled from map measurements. Figs. 30
and 31 deal with a medium-sized city, Fig. 30 predicat-
ing the exchange in the center, and Fig. 31 on the outside. The entire exchange is taken at 20,000 stations,
the average originating message rate for the whole at 16,
with office rates of 16 for Fig. 30, and 8 for Fig. 31. The
value of Q is taken from 100 to 150 for Fig. 30, and from
80 to 130 for Fig. 31, while q is allowed to vary from 65
per cent to 85 per cent. In Fig. 32 the whole exchange
is estimated at 10,000, the originating message rate at 10,
Q from 80 to 130 and q from 65 to 85 per cent. As in
Fig. 28 the average trunking distance is in all cases ascertained from map measurements.

The preceding data convey general average information
appertaining to outgoing trunk business, but it must not
be forgotten that the curves apply to trunking in one
direction only, and that the quantities must be doubled,
on the average, if both incoming and outgoing business is
to be cared for. Further, the curve quantities are *linear*

measurements, or *circuit* mileage, and must again be doubled if wire mileage is desired.

To accurately and completely design an interlacing Reverse Call Circuit Trunking System, and to determine that point at which the increasing trunk mileage overbalances the decreasing sub-station mileage, requires that the territory to be served be districted into a number of different offices, the telephonic centers located

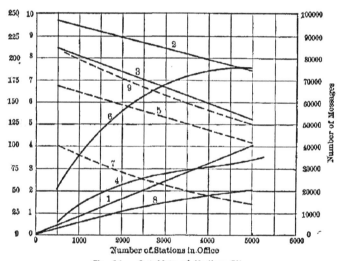

Fig. 31.— Outskirts of Medium City.

with some considerable care, and the trunk line routes settled and measured with a fair degree of accuracy, and then the sub-station and trunk mileage plotted for each of the different number of offices. In Fig. 27 a curve was given showing for an imaginary though typical large exchange, the rate of variation of the average sub-station lineage with the variation in the number of offices into which the territory was divided. If the total trunk lineage be divided by the sum of all the sub-stations the exact sub-station trunk lineage is obtained, and if this is done for each of the different number of

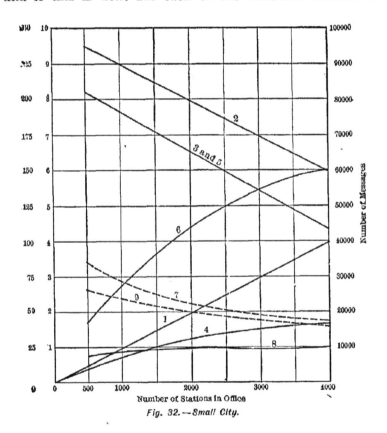

Fig. 32.—Small City.

offices into which the territory is sub-divided, data for plotting a curve that shows the rate of variation of trunk lineage with a change in the number of offices is obtained. In Fig. 33 curve C is plotted showing the change in trunk lineage per sub-station for the same conditions that were used in plotting the curves for sub-station lineage in Fig. 27, which are here repeated. Now by plotting a curve whose ordinates are the sum of the sub-station lineage plus the sub-station trunk lineage, while the abcissæ are the different number of offices into which the territory is divided, the total variation of wire mileage with a change in the number of offices is obtained, and by inspection it is easy to discover the least ordinate to this curve

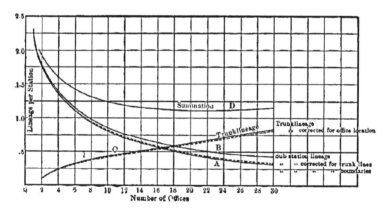

Fig. 33. — Lineage per Station.

and to determine that number of offices which will require the minimum wire plant. In the illustration about seventeen offices show the least ordinate, amounting to 1.2 miles, made up of .60 miles for the sub-station line, and .60 miles for the trunk-line mileage. Incidentally it is obvious from the discussion of this subject that the curve for sub-station mileage and trunk-line mileage per sub-station will intersect at the point of maximum economy in wire mileage, and that this point, like many other similar problems, is reached when the sub-station mileage and trunk mileage per sub-station are equal.

The trunk lines form a series of wires extending from every office in each group to all other offices, and evidently there will be trunk lines extending in all directions rectangularly to each adjacent office. In all but very exceptional cases the number of lines on the different sides of each office will be unequal. Thus, offices located on the outskirts of a group will have lines on three sides only, and frequently only on one side. Offices between the circumference and the center will have more lines extending towards the center and less towards the outskirts. So the addition of the trunk lines increases the number of wires radiating from each telephonic center. It has previously been shown that the telephonic center is so located as to place an equal number of sub-stations on each side of each co-ordinate axis drawn through it. This is equivalent to saying that there are an equal number of wires entering from the office on all four sides. The addition of the trunk lines will usually cause an unbalance in the number of wires on one or more sides. If now the office be moved in the direction of the preponderance of the number of trunk lines, it is evident that all the lines, both of the sub-stations and of the trunks, will be shortened in the direction that the office moves and lengthened in the other. By the principles previously enunciated the number of sub-station lines are equal on each side of the office, and until some sub-stations are crossed this equality is unaffected by moving the office. When trunk lines are added, therefore, the office *should move* in the direction of the preponderance of trunk lines till such a point is reached (by crossing sub-stations) as to cause an equal number of lines to enter on both sides of each co-ordinate axis. So after the number of trunk lines have been calculated, and the routes between the various offices selected, inspection will determine whether the location of the various telephonic centers will be sensibly affected. This is easily done on the map of the territory, the various offices being slightly shifted until an equal number of wires enter each, on each side of each co-ordinate axis. By this shifting the sub-station mileage is somewhat increased, while the trunk mileage is decreased, but the sum of the two is the true minimum. In the center of large towns or cities this displacement is almost inappreciable, but on the outskirts it may become a source of considerable error unless taken into consideration. Any change in the office location will, of course, react on the curve for sub-station mileage, which must be correspondingly corrected, as shown by the dotted line near the curve *A* (Fig. 33).

All these corrections are shown by the dotted lines in close proximity to the curves *B* and *C*, while the summation curve *D* is plotted for the corrected values. It will be noticed that this curve is quite flat near its least ordinate, there being only a difference of .025 of a mile per station between fourteen, eighteen, or twenty-two offices, so that quite a liberal latitude may be exercised in the selection of the desired number of offices without seriously changing the total required wire mileage.

The complete analysis of the relations of office location, and the relation of the number of offices to the wire plant,

has now been exhibited, and the methods of ascertaining
the change in wire mileage with any variation either in
location or number of offices in any exchange. To determine relative economy it is now necessary to investi-
gate the methods of construction and determine the cost
of wire plant.

INDEX.

151

www.ingramcontent.com/pod-product-compliance
Lightning Source LLC
LaVergne TN
LVHW012202040326
832903LV00003B/76